BRICS and MICs: Implications for Global Agrarian Transformation

T0136206

The economic and political rise of the BRICS countries (Brazil, Russia, India, China and South Africa), and powerful middle-income countries (MICs) such as Argentina, Indonesia and Turkey, has far-reaching implications for global agrarian transformation. These countries are key sites of agricultural commodity production, distribution, circulation and consumption and are contributing to major shifts in the character of agro-food systems.

This comprehensive collection explores these issues through the lens of critical agrarian studies, which examine fundamental social change in, and in relation to, rural worlds. The authors explore key themes such as the processes of agrarian change associated with individual countries within the grouping, the role and impact of BRICS countries within their respective regions, the role of other MICs within these regions and the rising importance of MICs within global and regional agro-food systems. The book encompasses a wide variety of case studies, including the expansion of South African agrarian capital within Africa; Brazil as a regional agro-food power and its complex relationship with China, which has been investing heavily in Brazil; the role of BRICS and MICs in Bolivia's soy complex; crop booms within China; China's role in land deals in Southeast Asia; and Vietnamese investment in Cambodia.

This book will be of interest to students and researchers of critical agrarian studies, with a focus on BRICS and MICs. It was originally published as a special issue of the journal *Globalizations*.

Ben Cousins holds a DST/NRF Research Chair in Poverty, Land and Agrarian Studies at the University of the Western Cape, Cape Town, South Africa.

Saturnino M. Borras Jr. is Professor of Agrarian Studies at the International Institute of Social Studies, The Hague, the Netherlands; Adjunct Professor in the College of Humanities and Development Studies (COHD) at China Agricultural University (CAU), Beijing, PR China; and a Fellow of the Transnational Institute, Amsterdam, the Netherlands.

Sérgio Sauer is Professor in the Post Graduate Program on Environment and Rural Development in the Centre for Sustainable Development at the University of Brasilia, Brazil. He has a research scholarship from CNPq, Brasilia, Brazil.

Jingzhong Ye is Professor and Dean of the COHD at CAU, Beijing, PR China.

Rethinking Globalizations

Edited by Barry K. Gills, University of Helsinki, Finland and Kevin Gray, University of Sussex, UK

This series is designed to break new ground in the literature on globalization and its academic and popular understanding. Rather than perpetuating or simply reacting to the economic understanding of globalization, this series seeks to capture the term and broaden its meaning to encompass a wide range of issues and disciplines and convey a sense of alternative possibilities for the future.

Disintegrative Tendencies in Global Political Economy
Exits and Conflict
Heikki Patomäki

Environmental Security in Transnational Contexts
What Relevance for Regional Human Security Regimes?
Edited by Harlan Koff and Carmen Maganda

The Role of Religion in Struggles for Global Justice
Faith in Justice
Edited by Peter J. Smith, Katharina Glaab, Claudia Baumgart-Ochse and Elizabeth Smythe

Vivir Bien as an Alternative to Neoliberal Globalization
Can Indigenous Terminologies Decolonize the State?
Eija Ranta

Feminist Global Political Economies of the Everyday
Edited by Juanita Elias and Adrienne Roberts

The Politics of Destination in the 2030 Sustainable Development Goals
Leaving No-one Behind?
Edited by Clive Gabay and Suzan Ilcan

BRICS and MICs: Implications for Global Agrarian Transformation
Edited by Ben Cousins, Saturnino M. Borras Jr., Sérgio Sauer and Jingzhong Ye

For more information about this series, please visit: https://www.routledge.com/Rethinking-Globalizations/book-series/RG

BRICS and MICs: Implications for Global Agrarian Transformation

Edited by
**Ben Cousins, Saturnino M. Borras Jr.,
Sérgio Sauer and Jingzhong Ye**

LONDON AND NEW YORK

First published 2019
by Routledge
2 Park Square, Milton Park, Abingdon, Oxon, OX14 4RN

and by Routledge
52 Vanderbilt Avenue, New York, NY 10017

First issued in paperback 2020

Routledge is an imprint of the Taylor & Francis Group, an informa business

British Library Cataloguing-in-Publication Data
A catalogue record for this book is available from the British Library

ISBN 13: 978-0-367-66411-4 (pbk)
ISBN 13: 978-0-367-13289-7 (hbk)

Typeset in Minion Pro
by codeMantra

Publisher's Note
The publisher accepts responsibility for any inconsistencies that may have arisen during the conversion of this book from journal articles to book chapters, namely the possible inclusion of journal terminology.

Disclaimer
Every effort has been made to contact copyright holders for their permission to reprint material in this book. The publishers would be grateful to hear from any copyright holder who is not here acknowledged and will undertake to rectify any errors or omissions in future editions of this book.

Contents

CONTENTS

Citation Information

The chapters in this book were originally published in the journal *Globalizations*, volume 15, issue 1 (January 2018). When citing this material, please use the original page numbering for each article, as follows:

Chapter 6

The agrifood question and rural development dynamics in Brazil and China: towards a protective 'countermovement'
Fabiano Escher, Sergio Schneider and Jingzhong Ye
Globalizations, volume 15, issue 1 (January 2018) pp. 92–113

Chapter 7

Chinese land grabs in Brazil? Sinophobia and foreign investments in Brazilian soybean agribusiness
Gustavo de L. T. Oliveira
Globalizations, volume 15, issue 1 (January 2018) pp. 114–133

Chapter 8

Land control and crop booms inside *China: implications for how we think about the global land rush*
Saturnino M. Borras Jr., Juan Liu, Zhen Hu, Hua Li, Chunyu Wang, Yunan Xu, Jennifer C. Franco and Jingzhong Ye
Globalizations, volume 15, issue 1 (January 2018) pp. 134–151

Chapter 9

Holding corporations from middle countries accountable for human rights violations: a case study of the Vietnamese company investment in Cambodia
Ratha Thuon
Globalizations, volume 15, issue 1 (January 2018) pp. 152–167

Chapter 10

Framing China's role in global land deal trends: why Southeast Asia is key
Elyse N. Mills
Globalizations, volume 15, issue 1 (January 2018) pp. 168–177

For any permission-related enquiries please visit:
http://www.tandfonline.com/page/help/permissions

Notes on Contributors

Moisés V. Balestro is Associate Professor of Social Sciences in the Research Center and Graduate Program on the Americas at the University of Brasilia, Brazil and leads a research group on Comparative Studies in Economic Sociology.

Saturnino M. Borras Jr. is Professor of Agrarian Studies at the International Institute of Social Studies, The Hague, the Netherlands; Adjunct Professor in the College of Humanities and Development Studies (COHD) at China Agricultural University (CAU), Beijing, PR China; and a Fellow of the Transnational Institute, Amsterdam, the Netherlands.

Ben Cousins holds a DST/NRF Research Chair in Poverty, Land and Agrarian Studies at the University of the Western Cape, Cape Town, South Africa.

Clara Craviotti is a Research Fellow of the National Scientific and Technical Research Council, Buenos Aires, Argentina and Professor at FLACSO Argentina, Buenos Aires, Argentina.

Fabiano Escher is a Postdoctoral Researcher in the Development, Agriculture and Society Graduate Program at the Federal Rural University of Rio de Janeiro, Brazil and at the National Institute of Science and Technology in Public Policies, Strategies and Development, Rio de Janeiro, Brazil.

Jennifer C. Franco is Adjunct Professor in the COHD at CAU, Beijing, PR China and a Researcher in the Agrarian and Environmental Justice Program, at TNI, Amsterdam, the Netherlands.

Ruth Hall is Professor in PLAAS, at the University of the Western Cape, Cape Town, South Africa.

Zhen Hu is Associate Professor in the COHD at CAU, Beijing, PR China.

Hua Li is Lecturer in the College of Politics and Law at Taiyuan University of Technology, PR China.

Juan Liu is Assistant Professor in the College of Humanities and Social Development at Northwest A&F University, Yangling, PR China and a Postdoctoral Researcher in ICTA at the Universitat Autónoma de Barcelona, Spain.

Ben M. McKay is Assistant Professor of Development and Sustainability at the University of Calgary, Canada.

Elyse N. Mills is a PhD Researcher at the ISS, The Hague, the Netherlands.

Gustavo de L. T. Oliveira is Assistant Professor of International Studies at the University of California, Irvine, USA.

Sérgio Sauer is Professor in the Post Graduate Program on Environment and Rural Development in the Centre for Sustainable Development at the University of Brasilia, Brazil. He has a research scholarship from CNPq, Brasilia, Brazil.

Sergio Schneider is Professor of Rural Sociology and Development in the Rural Development (PGDR) and Sociology (PPGS) Graduate Programmes at the Federal University of Rio Grande do Sul, Porto Alegre, Brazil.

Ratha Thuon is a PhD Candidate in the ISS at Erasmus University Rotterdam, The Hague, the Netherlands.

Chunyu Wang is Associate Professor in the COHD at CAU, Beijing, PR China.

Yunan Xu is a PhD Candidate in the ISS at Erasmus University Rotterdam, the Netherlands, funded by the China Scholarship Council.

Jingzhong Ye is Professor and Dean of the COHD at CAU, Beijing, PR China.

BRICS, middle-income countries (MICs), and global agrarian transformations: internal dynamics, regional trends, and international implications

Ben Cousins, Saturnino M. Borras Jr., Sérgio Sauer and Jingzhong Ye

ABSTRACT

The BRICS (Brazil, Russia, India, China, and South Africa) countries are emerging as key sites of agricultural commodity production, distribution, circulation, and consumption, contributing to major shifts in the character of regional and global agro-food systems. Their growing importance within the world food economy presents new challenges for scholars, activists, policy-makers, and development practitioners. The articles in this collection are located in their wider context, and the significance of their insights for a longer term research agenda within critical agrarian studies is explored. Four key themes are discussed: processes of agrarian change under way within BRICS countries; the role and impacts of BRICS countries in their respective regions; the rising importance of middle-income countries (MICs) within global and regional agro-food systems; and how the recent emergence of forms of populism, authoritarianism, and combinations of these two (i.e. 'authoritarian populism') is linked to the rise of the BRICS.

Introduction: framing a research agenda for critical agrarian studies

The economic and political rise of the BRICS (Brazil, Russia, India, China and South Africa) countries and powerful middle-income countries (MICs) such as Argentina, Chile, Indonesia, Malaysia, Nigeria, and Turkey, more or less from the early 1990s onwards, has far-reaching implications for global agrarian transformation. These countries are emerging as key sites of commodity production, distribution, circulation, and consumption, including in relation to agricultural commodities, and are contributing to major shifts in the character of regional and global agro-food systems. The five BRICS countries are working both separately, and increasingly together, to shape international development agendas, both as partners in and perhaps as an alternative to the mainstream development paradigms promoted by the traditional hubs of global capital and western-dominated international financial institutions such as the World Bank .

The rise of the BRICS countries alongside some powerful MICs, and emerging alliances between them, has sparked debates about whether or not they herald a new era for international economy and politics. Do they constitute an alternative to the conventional North Atlantic-anchored neoliberal prescription for capitalist development, or are their models of development problematic in both old and new ways (Bond & Garcia, 2015; Scoones, Amanor, Favareto, & Qi, 2016; Taylor, 2014)? More profoundly, the growing importance of BRICS countries within the world economy challenges dominant conceptions of global inequality, in which the North–South divide is viewed the most significant axis of major differences in power and wealth. What appears to be an evolving polycentric world order presents new challenges to, as well as opportunities for, scholars, activists, policymakers, development practitioners, and other actors, and processes of knowledge production need to respond to these.

In this introduction, we locate the articles in this special issue in their wider context, and explore the implications of the rise of the BRICS countries for critical agrarian studies. The latter focuses on questions of fundamental social change in relation to rural worlds, including the role of unequal power relations among and between agrarian classes and other social groups.

Constructing a research agenda for critical agrarian studies of the BRICS has to build on the insights of what other scholars have contributed to understanding the dynamics and impacts of this grouping of countries, especially those located in the disciplines of international political economy, international relations, international development studies, strategic, and geopolitical studies. A critical agrarian studies focus is urgent and necessary, in our view, because the rural dimensions of the BRICS are a strategically important component of these emerging realities, but are somewhat under-studied at present. It is true that academic research on the role of agriculture in bilateral relations between BRICS countries has blossomed in recent years, but most of these initiatives have been somewhat Africa-centric to date, and have tended to focus on the impact of development policies pursued by a specific BRICS country in African contexts: China in Africa, Brazil in Africa, and so on (Amanor & Chichava, 2016; Cabral, Favareto, Mukwereza, & Amanor, 2016; Hall, Scoones, & Tsikata, 2015; Scoones et al., 2016; Scoones, Cabral, & Tugendhat, 2013). This work has produced important insights, for example, on the influence of a country's history of agrarian development on how it frames programmes of developmental assistance to African countries. In attempting to construct a wider research agenda, it is important to build on high-quality scholarship of this kind – but also to expand the focus of work, as required.[1]

Key themes: agrarian change in BRICS countries

Much current scholarly research on the rise of the BRICS and MICs, while focusing on critically important themes such as impacts on the character of the global political economy, has left key gaps in relation to agrarian change, which the current collection attempts to address. It is important to note some key features of this work. First, the primary concern of these authors is not the nature and impacts of BRICS as an organization, but rather processes associated with individual countries within the grouping, both *within* their national borders and in their regional contexts. Second, although this collection does focus on the role and impacts of BRICS countries in other regions, such as investments by Chinese companies in Brazil (i.e. on *inter-regional* dynamics), equally important, but generally absent in the literature, is a key focus on BRICS countries in their respective regions (i.e. on *intra-regional* dynamics). Third, as a result of this intra-regional lens, authors also focus on the role of other important players within the respective regions, namely, MICs. The rising importance of MICs within global and regional agro-food systems is another under-explored dimension of contemporary agrarian transformations. These are evident in differences between the expanding activities and impacts of China in Southeast Asia vis-à-vis those of Thailand in East and Southeast Asia, and the complex role of Argentinian soya seed firms in the Brazilian seed industry. Fourth, the rise of different forms of populism, authoritarianism, and combinations of these two (i.e. 'authoritarian populism') is one of the defining features of the current political moment among BRICS countries, as well as many of the MICs, and key developed countries such as the United States and many countries in Europe. Our hunch is that the rise of populism, authoritarianism, or authoritarian populism in various parts of the world is interconnected, and that their rural roots and dynamics, although largely under-studied to date, need to be better understood (Scoones et al., 2018).

This collection, together with the recent set of papers published in *Third World Thematics*, can to be seen as a preliminary attempt to help kick-start a conversation around these themes, and to shape an emerging research agenda. Below we argue why such a research agenda matters for critical agrarian studies.

Internal agrarian transformations within BRICS countries

In each of the BRICS countries, profound changes are under way in rural society and their agrarian economies. These include increased levels of concentration in landholdings, changes in the character of rural–urban links, shifting patterns of migration, the promotion of smallholder farming alongside the rise of corporate agribusiness, increasing degrees of vertical integration within value chains, the intensified 'supermarketization' of food retailing, and different combinations of these processes. Understanding why BRICS countries have aggressively crossed borders to seize and take control of natural resources (land, water, seas, forests, minerals, commodity chain, and so on) in distant places requires getting to grips with the concrete conditions found in these settings, as well as identifying accumulation imperatives within the BRICS countries. These play out in various interrelated ways, including: (a) an over-accumulation of capital and hence the need to invest it elsewhere; this is evident in the Belt and Road Initiative of China; (b) the need to secure cheaper sources of the means of agrarian production (land, labour, raw materials); this is largely what has been driving the global land/resource rush that has involved the BRICS countries as farmland investors; (c) the limits of domestic markets within the BRICS countries, and the need to gain control of lucrative markets abroad; and (d) more straightforward political motivations, such as the need to secure a stable supply of cheap food for internal consumption, to appease an inherently politically volatile working class and ensure affordable food provisioning.

In addition, and closely linked to the above processes, a fuller understanding of why the BRICS countries have engaged in aggressive cross-border economic and political activities over the past two decades requires a systematic understanding of internal agrarian transformations within these countries. This in turn requires us to examine BRICS countries as key sites of the contemporary commodification of remaining agrarian commons' (land, water, seas, forest, and minerals) and labour, intimately related to the extension of the broader spheres and structures of commodity production, distribution, circulation, and consumption – and how these are dialectically linked to 'external' social, economic, and political processes. This is evident in a number of ways, two of which are strategically important.

On the one hand, we see a significant restructuring of commodity production, distribution, and circulation dynamics within these countries and beyond. For instance, the demise of soya production inside China was not an accident. Considerations of productivity and competitiveness motivated a government policy decision to aggressively outsource soya production (Hairong, Yiyuan, & Bun, 2016), thus triggering the rise of a BRICS country, Brazil – in combination with an MIC, Argentina, as a new global hub of soya capitalism, with far-reaching implications for the United States, the traditional site of most soya production and consumption (Oliveira, 2016, 2017; Oliveira & Schneider, 2016). Ripple effects included the emergence of a soya complex in the southern cone of Latin

America (Escher, Schneider, & Ye, 2017; Sauer, Balestro, & Schneider, 2017). McKay (2017) has shown how the global-regional restructuring of the soya complex in turn prompted the creation of the Bolivian soya complex, facilitated by both the Brazilian and Bolivian states and dominated by Brazilian capital.[2] Similarly, Craviotti (2017) discusses how Argentinean soya seed companies have responded to these processes by becoming *multilatin* firms, merging or joining (multi-) national companies taking advantage of governmental support (Sauer & Mészáros, 2017).

On the other hand, we see significant global shifts in relation to specific commodities that in turn trigger the restructuring of commodity chains within BRICS countries. For example, the rise of the 'flex sugarcane complex', in which commodities produced from sugarcane crop can be used for multiple purposes (as sweeteners, ethanol, and other commercial and industrial uses), has seen a net increase in the area of sugarcane area planted globally over the past two decades. Brazil is a key player in the global flex sugarcane complex (McKay, Sauer, Richardson, & Herre, 2016; Borras, Franco, Isakson, Levidow & Vervest, 2016), which is shaping the sugar industry in regions such as Southern Africa (Dubb, Scoones, & Woodhouse, 2017) and in countries such as India and China. Borras et al. (2017) trace the trajectory of the boom of sugarcane production in south-eastern China during the past decade or so, that has significantly different features from crop booms elsewhere. This is partly because it is based on production on the small plots of thousands of smallholders, who have either leased their lands to companies or have engaged directly in sugarcane production themselves. Unlike the soya sector, signals from the sugarcane sector indicated that production in south-eastern China could be globally competitive. However, recent indications suggest that the inability of Chinese producers to secure productivity increases through mechanization and irrigation may undermine the sustainability of the sugarcane boom, forcing companies to cross borders and tap into the cheaper means of production and labour in neighbouring Southeast Asian countries, as argued by Mills (2017) and Schoenberger, Hall, and Vandergeest (2017).

Interconnected processes of social change within and in relations between the agrarian sectors of BRICS countries, MICs and other countries, and the broader patterns they reveal, require deeper empirical investigation. Contributions to the current collection, together those in the special issue of *Third World Thematics* (2016), have been able to undertake only initial explorations. Many relevant and consequential questions require further reflection and investigation, including: what are the key similarities and differences between the agrarian structures of Brazil, Russia, India, China, and South Africa? What are the historical processes that shaped these countries' current situation, with special attention to convergences and divergences between these countries? What is the role of smallholders and family farmers in the agrarian system? What challenges and pressures are they confronted with, and what patterns of social differentiation are emerging? What forms of collective action are evident and possible? How are local and national processes of agrarian transformation being shaped by global and transnational processes of investment, trade, and inter-state relations (and vice versa)? What contradictions and antagonisms have emerged, and, as Escher et al. (2017) ask, how should social struggles in these contexts be characterized (e.g. in terms of Polanyi's notion of a 'double-movement')?

The role of BRICS countries within their regions

Internal agrarian transformations in BRICS countries are in turn interacting with changes in rural societies and agrarian economies in their neighbouring countries. As several articles in this collection discuss, BRICS countries are expanding their presence in their respective regions, partly by promoting state and corporate partnerships and investment deals, as well as supporting private individual business transactions. These processes do not simply represent the expansion of these countries into their respective regions, as 'imperial or sub-imperial' powers (Bond, 2015); rather, the strategies and actions of both states and companies interact with dynamic changes already under way within these regions. At least three important social processes are underscored by articles in this collection.

First, the cost of the means of production, principally land and labour, could, and did, rise to a level that rendered production inside BRICS countries relatively less competitive regionally or globally. This forced them to try to secure access to and control over cheaper land and labour supplies elsewhere, and their neighbouring countries have offered vast opportunities. When labour became scarce and expensive in the production of sugarcane in southern China, it was possible to maintain the level of crop production only by employing cheap Vietnamese migrant workers (Le, Vu, & Borras, 2017). In recent years, between November and March, up to 80,000 Vietnamese migrant workers per annum have illegally crossed the border into China to seek employment cutting cane. When land becomes scarce and expensive, and production costs in general rise in relation to output, expanding into the wider region becomes the most logical step to secure profitability, as demonstrated by South African sugarcane corporate giants in recent years (Dubb, 2017; Martiniello, 2016).

Second, the availability of both finance capital and relatively more lucrative investment opportunities elsewhere often entice companies from BRICS countries to cross borders. Available capital may either be that accumulated within the BRICS countries themselves, or originate elsewhere, and then invested in (or alternatively, passed through) a BRICS country. A dollar invested by a South African company in farmland in Zambia may yield better profits than when invested inside South Africa, for various reasons. That dollar may be drawn from over-accumulated capital held by a South African company, or it may be held by a German bank wanting to invest in farmland, but for a variety of reasons prefers is invested via a South African intermediary. Hall and Cousins (2017), Campbell (2016), and Martiniello (2016) have explored these dynamics quite extensively in the African context.[3] We see similar regional dynamics elsewhere. They are at the core of what is known as Trans-Latina Corporations or *multilatin* firms (Borras, Franco, Gómez, Kay, & Spoor, 2012; Craviotti, 2017; Sauer et al., 2017). In these processes, the ultimate provenance of much the capital that is invested, although apparently originating from the Cayman Islands or Panama, is often Europe, Russia, China, Brazil, or the United States (Borras et al., 2012).

Third, the combination of increasingly vibrant regional economic hubs and relatively loose regulation of capital facilitates the emergence of layers of individual entrepreneurs, speculators, brokers, and scammers straddling increasingly porous national borders. The past decade, for example, has seen an increase in farmland investments but without the visible involvement of well-known, iconic corporations. Many of these are 'stealth transactions', only some of which are legal. The rise of the soya complex in Paraguay and in Bolivia, for example, has been anchored not by iconic Brazilian soya corporations, but by a migrating and enterprising mass of *gauchos* – rich southern Brazilian farmers of European descent. Some of them are already citizens of Paraguay or Bolivia, others are not, but they remain firmly linked to Brazil and are even encouraged by the Brazilian state (McKay, 2017; Sauer & Mészáros, 2017; Wilkinson, Valdemar, & Lopane, 2016). There has been a similar process involving individual white commercial farmers from South Africa moving north in to the rest of Africa (Hall, 2012; Hall & Cousins, 2017). This kind of process is also responsible for the profound transformation of land holdings in the northern states of Myanmar (Franco & Borras, in press). It can involve areas of farmland larger than those openly acquired by corporations, and is difficult to govern because it operates below the radar of formal regulatory institutions.

In this collection, authors have managed to only scratch the surface of these kinds of intra-regional dynamics. Many questions remain, including the following: what kinds of agrarian transformation are underway in the regions in which BRICS countries are located? What are the dominant (but also countervailing) directions of change in agro-food systems, and what endogenous and exogenous forces are driving such changes? What role is played by state policies, and how do these condition the behaviour of private sector actors – and in turn, how are private sector forces shaping state responses? What are the outcomes of changes in demographic structure, patterns of settlement and migration, input systems, levels investments in agriculture and other rural productive sectors, levels of agricultural production and the rise of boom crops, agro-processing and the rise of manufactured foods, and the relative importance of formal and informal markets for agricultural products? What is the character of relations between the BRICS countries and other MICs located within their regions? What food system is emerging at the regional level, who are the winners and losers in this, and where is this spatially located? What are the changing dynamics of the patterns of production, circulation, and consumption of food, at national, regional, and global levels? What factors – including political, demographic, economic, environmental and climatic – are driving these changes?

In addition, we can ask: in what ways are agricultural and food value chains being restructured? What are the roles of, and interests shaping, the retail sector's expansion and what is the scope, significance, and politics of its promotion of alternative food production and restructured value chains (organic, buying local, preferential procurement)? How is concentration of ownership and control in these value chains occurring, and how have power relations shifted between family farmers and agribusiness sectors, and among them? With what effects on production and accumulation? What is the role of BRICS countries in the emerging financialization of agriculture? How are the BRICS countries' roles in global trade in food and agricultural commodities changing? What are their histories of trade and trading partners, and through what historical processes have they helped to constitute the current global agro-food system? Through what political, legal, and other processes are the current changes unfolding? How is membership of the BRICS group influencing and enabling this change, and how are other actors responding, and to what effect? What new sites of contestation around food production, circulation, and consumption are opening up and what possibilities for alternatives are emerging?

The rise of MICs in relation to BRICS

What makes the rise of the BRICS countries both interesting and complex is the parallel rise of a small number of powerful MICs that have become important sites of agricultural commodity production, distribution, circulation, and consumption in their own right. Examples include Chile and Argentina in Latin America; Thailand, Malaysia, and Vietnam in Southeast Asia; and Nigeria and Kenya in Africa. These countries are also sites of important internal agrarian transformations over the past few decades, marked by deepening commodification of their natural resources and labour. Articles in this collection contribute two key insights.

First, BRICS countries are in either alliance or competition with other regional powers. Agribusiness, mining, food, real estate, finance, and banking corporations from MICs have also become large and powerful in their domestic markets, and hence have begun to look beyond their national borders for opportunities to engage in expanded accumulation (Campbell, 2016; Hall & Cousins, 2017; McKay, 2017). From a regional perspective, corporations from MICs are just as pervasive, entrenched and influential as their counterparts from both BRICS countries and traditional hubs of capital. In some instances, companies from BRICS countries forge alliances with these other economic powers, on other occasions they compete with them. Chinese and Vietnamese companies, for example, are among the most dominant farmland and mining investors in Cambodia and Laos, and they appear to coordinate decisions on which corporation will acquire which lands in these countries, facilitated by supportive and friendly governments (Thuon, 2017). In the sugarcane boom and eucalyptus tree plantation expansion in south-eastern China, foreign companies, including Thai sugarcane companies, and an Indonesian pulp and paper company, compete fiercely with Chinese companies (Borras et al., 2017).

Second, companies in MICs also aggressively cross national borders, including into BRICS countries, in order to secure access to cheap raw materials, land and labour (often cheaper than their own), or because the targeted country has a more lucrative market than their own home market. In doing so, they have become either coveted investors or despised land grabbers in neighbouring countries. They tend to target two kinds of investment destination: (i) poorer neighbouring countries and (ii) equally powerful and wealthy countries in their regions. Thuon (2017) describes why and how a Vietnamese company seized control of land in Cambodia for rubber production, generating much conflict with Cambodian villagers.

This is not an isolated case. Vietnamese corporations are active in many countries, and entrenched in Cambodia and Laos. A Malaysian oil palm company has seized control 48,000 acres of Karen villagers' land for oil palm plantation (Franco & Borras, in press). Thai sugar companies are among the largest and most dominant corporate players in the sugarcane boom in south-eastern China (Borras et al., 2017). One of the foreign paper and pulp companies that has seized control of an enormous amount of villagers' land in China in order to produce eucalyptus is Indonesian (Xu, 2018a, 2018b). Argentinean soya companies are among the most significant investors in Paraguay, while Chilean timber companies are among the biggest firms operating in Brazil (Andrade, 2016).

In the context of the global land rush literature and the many debates it has generated, it is clear that: (a) the role of MICs is often missed and (b) the reality is that BRICS countries – often portrayed as the domestic base of large-scale land grabbers – are actually also important sites of land grabbing by foreign companies. This complicates how we think about agro-commodity booms and global commodity chains, and these insights should inform both strategies to resist land grabs and attempts to formulate alternative approaches to agro-food system governance.

Populism, authoritarianism, and authoritarian populism

The rise of the BRICS countries and other MICs has coincided with the rise of populism, authoritarianism, and authoritarian populism, worldwide. Is this a random occurrence, or is there a link between these two phenomena? Our hunch is that there is a connection between the two, and this can be seen in a number of ways.

First, the economic rise of BRICS countries and MICs was partly the result of manufacturing capital migrating from industrialized countries to the Global South, seeking greater profits by securing cheaper raw materials, cheaper labour, fewer environmental regulations, reduced taxes, and access to lucrative markets. As industrial jobs multiplied in the BRICS and MICs, they began to disappear in parts of the Global North, resulting in abandoned industrial belts such as the so-called Rust Bel' in the mid-West of the United States, or in zones marked by the absence of jobs, enterprises and economic opportunities, such as rural America or the north of the UK. This has in turn contributed to the rural population's support for right wing, nationalist and xenophobic forms of politics that have capitalized on the issues of rural marginalization and powerlessness. This helped to get Trump elected in the United States, and nearly led to the election of Marine Le Pen in France, for instance (Scoones et al., 2018; Ulrich-Schad & Duncan, 2018).

Second, the rise of right wing, nationalist, and xenophobic political movements in several industrialized countries, e.g. United States, France, Germany, and Austria, have generated forms of public rhetoric in which BRICS countries and MICs are accused of 'stealing' jobs and factories from citizens of 'industrialized' countries. In turn, this has provoked right wing and nationalist posturing and rhetoric from leaders in some of these countries, e.g. the Hun Sen regime in Cambodia and the Duterte government in the Philippines, with negative socio-economic and political consequences for rural inhabitants in these countries.

Third, the increased traffic of (usually illegal) migrant farmworkers and other members of the rural poor across national borders, such as Morocco – Italy/Spain, Mexico/Central America – USA/Canada, Zimbabwe/Mozambique – South Africa, and Vietnam – south-eastern China, has further contributed to the rise of xenophobic and nationalist-populist tendencies in many countries, even when the rural jobs they take up are unlikely to be filled by citizens of host countries.

In short, it is possible that the rise of various strands and combinations of populism and authoritarianism within the BRICS and MICs is intimately linked to the rise of similar political tendencies in wealthy countries in the Global North. Articles in this collection have not tackled this issue to any significant extent. Yet, we argue that this is an important nexus of questions that require broad-ranging empirical inquiry and theorizing within the field of critical agrarian studies and beyond.

Implications for critical agrarian studies more broadly

Individually and collectively, the contributions to this collection offer multiple insights for contemporary critical agrarian studies in general, on the one hand, and for our efforts at developing a fuller understanding of the causes and conditions of the rise of the BRICS countries and MICs, and the consequences for global agrarian transformations, on the other hand. We conclude by discussing a few key themes.

First, the agrarian dimension of the rise of BRICS countries is critically important. Processes of agrarian change in these countries and in their neighbouring regions involve an extremely large number of rural people and an enormous quantity of natural resources located inside these counties and elsewhere in the world. Yet, this has been one of the most under-explored dimensions of scientific research and public debates on the BRICS. The available studies that engage with agrarian issues tend to be focused on how BRICS countries secure natural resources and markets in distant regions – involving inter-regional dynamics (e.g. McKay, Alonso-Fradejas, Brent, Sauer, & Xu, 2016 for China in Latin America), relations between BRICS countries (e.g. Zhou, 2017, for Chinese investments in Russia), and so-called South-South links (e.g. Milhorance, 2017, for Brazil in Latin America and Africa). This collection validates the importance of these, and at the same time demonstrates that the full range of agrarian dynamics is in fact far more extensive and complex, and requires systematic research. Our current collection, together with the special issue of *Third World Thematics*, has made only preliminary contributions to a deeper understanding of the BRICS phenomenon, perhaps provoking more questions than providing definitive answers.

Second, the parallel rise of MICs requires deeper examination. While an increasing number of empirical studies suggest that MICs have been transformed into key sites of commodity production, distribution, circulation, and consumption, none have examined such dynamics in a generalized manner or in relation to the rise of BRICS countries. We have barely scratched the surface of this issue in our own collection, but it does provide us with the basis for suggesting that this particular dimension of global agrarian transformation warrants deeper and more systematic empirical investigation and theorizing.

Third, multi-directional processes of change in social relations around land, labour, and capital within and between BRICS countries are most concretely observed when examined at a regional level, where the role of MICs is also most concrete. While there are quite a number of regionally situated studies in the context of contemporary agrarian transformations (e.g. Schoenberger et al., 2017), there have been few systematic empirical and theoretical studies of specific regional sites as a unit of analysis. This collection includes some interesting and provocative individual contributions on particular regions, e.g. Sauer et al. (2017) and Hall and Cousins (2017), but at best these allow us to argue that deeper and more systematic inquiry along these lines has great potential.

Fourth, this special issue demonstrates that the importance of current discussions about global governance of the agri-commodity boom and the global resource rush is both re-affirmed and questioned. Re-affirmed, because many of the global governance instruments being discussed at present clearly have great potential if interpreted and activated more universally. Questioned, because most discussions of these global governance instruments tend to be focused only on lower income countries in the Global South, rather than more broadly, but are highly relevant in MICs too (see related discussion by Franco, Park, & Herre, 2017).

Finally, we suggest that discussions of populism, authoritarianism, or authoritarian populism can be deepened and broadened if seen in the context of the rise of the BRICS countries and MICs. It is not an accident that the BRICS countries and many of the MICs are sites of the emergence of these forms of politics. They are not isolated from the rise of populism, authoritarianism, or authoritarian populism across the globe, but rather closely connected to the phenomenon in various ways and with a variety of implications. These 'various ways' and 'variety of implications' warrant deeper and more systematic empirical investigation if we want to better understand these kinds of politics – and how they link to the rural world.

Notes

1. A recent special issue of *Third World Thematics* (McKay et al., 2016 and others) contains an important set of contributions on processes of agrarian change in BRICS countries and their regions, and these are further explored in this collection. The articles in both collections arise from work undertaken for the BRICS Initiative for Critical Agrarian Studies (BICAS) – see www.iss.nl/bicas and www.plaas.org.za/bicas. Most of the scholars participating in this initiative are located in BRICS countries.
2. For a broader perspective on this specific example, see Wilkinson et al. (2016).
3. See Campbell (2016) for why South African supermarkets are investing heavily in other Southern African countries.

Acknowledgements

Many thanks to the editor of *Globalizations*, Barry Gills, for his guidance throughout the editing of this special issue.

Disclosure statement

No potential conflict of interest was reported by the authors.

Funding

The work for this collection was supported in part by the Ford Foundation's Beijing Office (grant number 0135-1532-0).

References

Amanor, K. S., & Chichava, S. (2016). South–south cooperation, agribusiness, and African agricultural development: Brazil and China in Ghana and Mozambique. *World Development, 81*, 13–23.

Andrade, D. (2016). 'Export or die': The rise of Brazil as an agribusiness powerhouse. *Third World Thematics: A TWQ Journal, 1*(5), 653–672.

Bond, P. (2015). BRICS and the sub-imperial location. In P. Bond & A. Garcia (Eds.), *BRICS: An anti-capitalist critique* (pp. 15–26). Johannesburg: Jacana Media.

Bond, P., & Garcia, A. (Eds.). (2015). *BRICS: An anti-capitalist critique*. Johannesburg: Jacana Media.

Borras, S. M., Jr., Franco, J. C., Gómez, S., Kay, C., & Spoor, M. (2012). Land grabbing in Latin America and the Caribbean. *Journal of Peasant Studies, 39*(3–4), 845–872.

Borras, S. M., Franco, J. C., Isakson, S. R., Levidow, L., & Vervest, P. (2016). The rise of flex crops and commodities: Implications for research. *Journal of Peasant Studies, 43*(1), 93–115.

Borras, S. M., Liu, J., Hu, Z., Li, H., Wang, C., Xu, Y., … Ye, J. (2017). Land control and crop booms inside China: Implications for how we think about the global land rush. *Globalizations*. doi:10.1080/14747731.2017.1408287

Cabral, L., Favareto, A., Mukwereza, L., & Amanor, K. (2016). Brazil's agricultural politics in Africa: More food international and the disputed meanings of 'family farming. *World Development, 81*, 47–60.

Campbell, M. (2016). South African supermarket expansion in sub-Saharan Africa. *Third World Thematics: A TWQ Journal, 1*(5), 709–725.

Craviotti, C. (2017). Agrarian trajectories in Argentina and Brazil: Multilatin seed firms and the South American soybean chain. *Globalizations*. doi:10.1080/14747731.2017.1370274

Dubb, A. (2017). Interrogating the logic of accumulation in the sugar sector in Southern Africa. *Journal of Southern African Studies, 43*(3), 471–499.

Dubb, A., Scoones, I., & Woodhouse, P. (2017). The political economy of sugar in Southern Africa – introduction. *Journal of Southern African Studies, 43*(3), 447–470.

Escher, F., Schneider, S., & Ye, J. (2017). The agrifood question and rural development dynamics in Brazil and China: Towards a protective 'countermovement'. *Globalizations*. doi:10.1080/14747731.2017.1373980

Franco, J. C., & Borras, S. M. (in press). Land politics in the era of climate change: Exploring the frontiers of global land grabbing. *Environmental Research Letters*.

Franco, J. C., Park, C., & Herre, R. (2017). Just standards: International regulatory instruments and social justice in complex resource conflicts. *Canadian Journal of Development Studies, 38*(3), 341–359.

Hairong, Y., Yiyuan, C., & Bun, K. H. (2016). China's soybean crisis: The logic of modernization and its discontents. *Journal of Peasant Studies, 43*(2), 373–395.

Hall, R. (2012). The next Great Trek? South African commercial farmers move north. *Journal of Peasant Studies, 39*(3-4), 823–843.

Hall, R., & Cousins, B. (2017). Exporting contradictions: The expansion of South African agrarian capital within Africa. *Globalizations*. doi:10.1080/14747731.2017.1408335

Hall, R., Scoones, I., & Tsikata, D. (Eds.). (2015). *Africa's land rush: Rural livelihoods and Agrarian change*. Woodbridge: James Currey.

Le, H., Vu, H., & Borras, S. M. (2017). *Vietnamese migrant labour and the rise and maintenance of the sugarcane boom inside China: Causes, conditions, consequences*. A paper presented at an international conference on 'BRICS and Agro-extractivism', Moscow, 13–16 October 2017.

Martiniello, G. (2016). 'Don't stop the mill': South African capital and agrarian change in Tanzania. *Third World Thematics: A TWQ Journal, 1*(5), 633–652.

McKay, B. (2017). Control grabbing and value-chain agriculture: BRICS, MICs and Bolivia's soy complex. *Globalizations*. doi:10.1080/14747731.2017.1374563

McKay, B., Alonso-Fradejas, A., Brent, Z., Sauer, S., & Xu, Y. (2016). China and Latin America: Towards a new consensus of resource control? *Third World Thematics: A TWQ Journal, 1*(5), 592–611.

McKay, B., Sauer, S., Richardson, B., & Herre, R. (2016). The political economy of sugarcane flexing: Initial insights from Brazil, Southern Africa and Cambodia. *Journal of Peasant Studies, 43*(1), 195–223.

Milhorance, C. (2017). Growing South-South agribusiness connections: Brazil's policy coalitions reach Southern Africa. *Third World Thematics: A TWQ Journal, 1*(5), 691–708.

Mills, E. (2017). Framing China's role in global land deal trends: Why Southeast Asia is key. *Globalizations*. doi:10.1080/14747731.2017.1400250

Oliveira, G. (2016). The geopolitics of Brazilian soybeans. *Journal of Peasant Studies, 43*(2), 348–372.

Oliveira, G. (2017). Chinese land grabs in Brazil? Sinophobia and foreign investments in Brazilian soybean agribusiness. *Globalizations*. doi:10.1080/14747731.2017.1377374

Oliveira, G., & Schneider, M. (2016). The politics of flexing soybeans: China, Brazil and global agro-industrial restructuring. *Journal of Peasant Studies, 43*(1), 167–194.

Sauer, S., Balestro, M. V., & Schneider, S. (2017). The ambiguous stance of Brazil as a regional power: Piloting a course between commodity-based surpluses and national development. *Globalizations*. doi:10.1080/14747731.2017.1400232

Sauer, S., & Mészáros, G. (2017). The political economy of land struggle in Brazil under workers' party governments. *Journal of Agrarian Change, 17*(2), 397–414.

Schoenberger, L., Hall, D., & Vandergeest, P. (2017). What happened when the land grab came to Southeast Asia? *Journal of Peasant Studies, 44*(4), 697–725.

Scoones, I., Amanor, K., Favareto, A., & Qi, G. (2016). A new politics of development cooperation? Chinese and Brazilian engagements in African agriculture. *World Development, 81*, 1–12.

Scoones, I., Cabral, L., & Tugendhat, H. (2013). New development encounters: China and Brazil in African agriculture. *IDS Bulletin, 44*(4), 1–19.

Scoones, I., Edelman, M., Borras,Jr.S. M., Hall, R., Wolford, W., & White, B. (2018). Emancipatory rural politics: Confronting authoritarian populism. *The Journal of Peasant Studies, 45*(1), 1–20.

Taylor, I. (2014). *Africa rising? BRICS - diversifying dependency*. London: Boydell & Brewer.

Thuon, R. (2017). Holding corporations from middle countries accountable for human rights violations: A case study of the Vietnamese company investment in Cambodia. *Globalizations*. doi:10.1080/14747731.2017.1370897

Ulrich-Schad, J. D., & Duncan, C. M. (2018). People and places left behind: Work, culture and politics in the rural United States. *Journal of Peasant Studies, 45*(1), 59–79.

Wilkinson, J. J., Valdemar, W. J., & Lopane, A. R. M. (2016). Brazil and China: The agribusiness connection in the southern cone context. *Third World Thematics: A TWQ Journal, 1*(5), 726–745.

Xu, Y. (2018a). Land grabbing involving foreign investors in China: Boom crops and rural conflict. Unpublished manuscript.

Xu, Y. (2018b). Loss or gain? Insights from Chinese villagers' livelihood change in the boom of eucalyptus trees. Unpublished manuscript.

Zhou, J. (2017). Chinese agrarian capitalism in the Russian far East. *Third World Thematics: A TWQ Journal, 1*(5), 612–632.

Exporting contradictions: the expansion of South African agrarian capital within Africa

Ruth Hall and Ben Cousins

ABSTRACT

Agrarian change in South Africa over the past two decades has seen consolidation of the hegemony of large-scale commercial farming and corporate agribusiness within agro-food systems. Constrained domestic demand and growth opportunities elsewhere have driven both farming and agribusiness capitals to move into other African countries, attempting to reproduce agro-food systems similarly centred on the dominance of large capital. This is evident in five areas: first, the financialization of agriculture and 'farmland funds'; second, multinational and South African input supply industries; third, large-scale land deals to expand industrial farming systems; fourth, the export of South African companies' food processing, manufacture, logistics and distribution operations; and fifth, the expanding reach of South African supermarkets and fast food chains. Regional expansion involves South African agrarian capital encountering substantial obstacles to entry, and challenges mounted by competitors in destination markets. Success as a regional hegemon in Africa's agro-food system is thus far from assured, and even where it does appear to succeed, generates contradictions, and rising social tensions of the kinds experienced in South Africa itself.

1. Introduction

Spurred by rapid deregulation and liberalization, the overall trajectory of agrarian change in South Africa over the past two decades has seen consolidation of the hegemony of large-scale commercial farming and corporate agribusiness within agricultural value chains. Ownership and control have become highly concentrated; high-tech and high-input production systems are focused on lucrative new crops and markets; and employment continues to decline. In a context of constrained domestic demand due to high levels of unemployment and poverty and stagnating growth, and emerging opportunities for geographic diversification, both farming and agribusiness capitals are now expanding into African countries. Their strategies are premised on promoting and reproducing agro-food systems centred on the dominance of large capital. However, the support of host states, of the kind that facilitated the large-scale form of capitalist production and the growth of agribusiness companies in South Africa, is not assured, and success is thus far from certain.

Large-scale commercial farmers and agribusiness corporations have been expanding their operations into the wider region in recent years, but so have many other South African companies (Boche & Anseeuw, 2013; Hall, 2012). Such expansion is driven by calculations of potential profitability, but is also shaped by specific conditions. These arise from a combination of the inherited structure of the economy at the end of apartheid, neoliberal economic policies adopted by the post-apartheid government after 1994, and the slowing down of the global capitalist economy since the financial crisis of 2008/2009 (Fine, 2008). But the strategies and *modi operandi* of South African companies were developed in a particular political economy context, and bear its imprint; when exported to other African countries, they tend to generate tensions reminiscent of those experienced in the domestic context.

In this article we map out some of the structural foundations underpinning the regional expansion of South African capital, identifying key features of the domestic agro-food economy which condition the behaviour of regionalizing companies. This is evident in five arenas: first, the financialization of agriculture and the emergence of South African-based 'farmland funds'; second, the growing influence of multinational and South African input supply industries; third, the prevalence of large-scale land deals premised on investment in agro-industrial farming systems; fourth, the expansion of South African companies' food processing, manufacture, logistics, and distribution operations; and fifth, the rapidly growing reach of South African supermarkets and fast food chains. In each arena, we identify key actors, noting their diverse expansion strategies and the contradictions they produce, including the ways in which these are circumscribed both by the character of local agro-food systems and by competition from other corporate investors in destination markets.

South Africa's regional expansion exemplifies a wider pattern of growing agricultural sector investment in regional and cross-regional agro-food systems by capitalist firms located in BRICS (Brazil, Russia, India, China, and South Africa) countries, and raises questions as to the similarities and differences among them (Scoones, Smalley, Hall, & Tsikata, 2014). Our focus in this article, however, is not with the BRICS organization per se, but rather with South Africa as one among several BRICS countries with growing influence in Africa's changing agro-food system. South Africa is, of course, an outlier among the group, having by far the smallest economy of those within the grouping. Its inclusion among the BRICS was widely perceived as being due to its economically and politically dominant position in Africa – rather than being an emerging global power. Yet even within Africa, South Africa's political and economic dominance is contested and increasingly in doubt (Radebe, 2016). Nonetheless South African companies are, together with others from within Africa and beyond, reshaping agro-food systems across parts of the continent in distinctive ways, albeit in a highly uneven and contradictory manner.

2. The political economy of post-apartheid South Africa

Capitalist development in South Africa took off towards the end of the nineteenth century, following the 'minerals revolution' – the discovery of diamonds in 1867 and gold in 1886, involving significant levels of foreign investment. Mining played a key role in driving urbanization and industrialization in the early twentieth century, and dominated export earnings for many decades, together with some agricultural products. Both mining and the (largely white) capitalist agriculture that it gave rise to depended on a low-wage regime. State policies in both the period of segregation following the consolidation a national polity ('union') in 1910 and in the apartheid era, from 1948 to 1994, focused in large part on ensuring a ready supply of cheap labour to these sectors (Marais, 2011).

The first democratic government of 1994 faced the challenge of transforming the economy to make it more equitable as well as being able to sustain growth. Freedom was achieved at the very moment that the global economy began to move towards ever higher levels of interaction and integration amongst national economies, led by the advanced capitalist countries and the USA in particular. Trade liberalization in South Africa after 1994 led to sharp increases in both export and imports, but globalization also brought many challenges, including those of intense competitive pressures from rapidly growing economies such as China.

Crucially, the manufacturing sector has not succeeded in becoming more dynamic, job-generating, and competitive in the post-apartheid era. The economy has shifted away from agriculture and levels and patterns of consumption have changed, but 'poverty is declining slowly, inequality is extremely high, and production and trade patterns have not shifted from the relative predominance of raw materials, exports and the importation of high value added manufactures' (Bhorat, Hirsch, Kanbur, & Ncube, 2014, p. 13). The economic prospects of the majority of citizens have been only slightly improved, notwithstanding the growth of a new black middle class and an extensive system of social grants that provides some relief for the poor.

Key features of the South African economy since the end of apartheid include the rapid expansion of services and government spending, and the continued relative decline of mining, manufacturing, and agriculture. Unemployment currently stands at around 27% when only active job seekers are counted, and at around 37% when those too discouraged to seek work are included. Many of those in employment, particularly in casual or temporary work, earn very low wages. More than half of the population is poor, and levels of inequality remain amongst the highest in the world, with a Gini coefficient that generally lies between 0.65 and 0.70 (Bhorat et al., 2014, p. 6). Poverty and inequality underlie severe social and political tensions, increasingly evident in relation to land, agriculture, and the agro-food system more broadly.

3. Continuity and change in South Africa's agro-food systems

Capitalist agriculture in South Africa and its dynamics must be understood within the larger structure and functioning of the country's political economy as a whole, as it has evolved over time, and its contradictions. South Africa's mining industry in the nineteenth and twentieth centuries was based on an authoritarian labour system designed to supply low-waged African migrant workers, as indicated above. These workers depended on food production by rural homesteads to supplement (i.e. subsidize) their wages, and were increasingly confined to densely populated 'native reserves', formed on the basis of large-scale dispossession of African land (Wolpe, 1972). A capitalist agriculture based on white-owned and generally large-scale farms emerged in response to rapidly growing markets in urban centres, and was strongly supported by the state, as well as by mining capital, for which cheap food formed a crucial 'wage good'. An alliance of 'gold and maize' was thus formed, centred on the Highveld in the interior, where most grain crops were produced. The development of capitalist agriculture in South Africa did not promote industrialization, as elsewhere, rather the reverse was true. The transition from pre-capitalist farming involved 'accumulation from above' on the basis of a 'Prussian-like' resolution of the agrarian question, as white land-owners, a key political constituency for Afrikaner nationalists, began to derive income from production rather than rent from African tenant farmers (Bernstein, 1996, p. 30).

These features involved high levels of state subsidy and support, in the form of cheap credit, guaranteed markets, and administered prices through marketing boards, tariffs, and other forms of protection, income support, infrastructure provision, and research and extension (Vink & Rooyen, 2009). They were accompanied by continuing repression of black workers, lack of support for black farmers, and forced removals aimed at achieving a racially segregated countryside. The rise to power of the National Party in 1948 and the adoption of apartheid as national policy saw intensified repression and adoption of measures such as the 'pass laws', aimed at heightened control of movement and settlement by the black majority.

By the 1970s many decades of intensive support had yielded a (white) capitalist agricultural sector characterized by increasing capitalization and mechanization, soaring output, but also farming systems premised on low wages and, increasingly, high levels of debt. By the 1980s the economic and political crisis of the apartheid regime saw increasing political pressure exerted by mining and industrial capital to reduce state support to agriculture and allow freer rein to 'market forces' (Bernstein, 1996, p. 31). The deregulation and liberalization of agriculture were initiated then, but came to fruition only after the transition to democracy in 1994.

Agricultural reforms since the transition have been largely detached from land reform, and have maintained their neoliberal character (Cousins, 2013). The concentration of ownership has proceeded apace, and by 2002 over half of all farm income was earned by 5% of enterprises (Vink & Rooyen, 2009, p. 32). Farm employment has fallen steadily, and is increasingly casualized and seasonal in nature. The number of large-scale commercial units is currently estimated at under 35,000. In contrast, a minority of black rural families earn cash from regular sale of agricultural produce – perhaps as few as 8% of black rural households with access to land, or between 200,000 and 400,000 production units (Aliber, Maluleke, Manenzhe, Paradza, & Cousins, 2013). Many of these supply informal local markets.

Post-1994 land reform policies aim to address the legacies of this bitter history of dispossession and discrimination through three mechanisms. A *land restitution* programme seeks to restore the land to black communities and individuals who were dispossessed in the period subsequent to the adoption of the Natives Land Act of 1913, which formally divided the entire country into race zones. The programme has experienced many problems, and has contributed little to reshaping ownership patterns in the countryside (Walker, Bohlin, Hall, & Kepe, 2010). *Land redistribution* aims to create a more equal distribution of land, but has slowed and instead of supporting black smallholders, in recent years has been diverted into tenuous forms of leasehold for a small grouping of aspirant black commercial farmers (Hall & Kepe, 2017). *Tenure reforms* were initiated to secure the legally vulnerable land rights of farmworkers and dwellers living on private farms as well as the 'customary' rights of people resident in the former reserves, now known as communal areas, but these show little evidence of success. The land tenure rights of the majority of black rural dwellers remain insecure (Cousins & Hall, 2013).

Ambitious land reform policies were adopted post-1994, but the ruling party has appeared to lack the political will required to tackle the complexities of the land question, displaying an enduring commitment to large-scale and commercial farming, despite rhetorical support for smallholder agriculture (Hall, Anseeuw, & Paradza, 2015). Over the past 22 years, funds for land reform have rarely constituted more than 0.5% of the national budget. The track record to date in delivering land and secure land rights to black South Africans is exceedingly poor, with only 8% of commercial farmland transferred (against a target of 30% by 2014), and many land reform projects performing badly in terms of production and the enhancement of rural livelihoods. Land reform is widely recognized as having failed to alter racially skewed inequalities in land holdings, and is thus a powerful symbol of the lack of substantive transformation in democratic South Africa (Cousins & Walker, 2015). The land is currently at the centre of fierce political contestation between and within political parties, and calls are being made to change the post-apartheid constitution to allow for expropriation of land without compensation. However, revelations of state capture by a politically well-connected family and their hangers-on are prompting anger at elite capture more widely (Bhorat et al., 2017).

Deregulation, especially the closure of marketing boards and privatization of sectoral cooperatives, has facilitated vertical integration in the agro-food system, leading to 'Big Food' – the growing dominance of just a handful of powerful corporations – in inputs, processing and retail, and to a lesser degree in primary production. This was accompanied by increased financialization, the key moment in which was the creation of a futures market (SAFEX) in key staple commodities. While financialization is global, in South Africa it was clearly a response to falling rates of profit, underpinned by liberalization, and South African capital responding to a wider set of opportunities and pressures in the global capitalist economy. The dismantling of the systems and institutions providing state support to agriculture has been accompanied by the growing power of agribusiness corporations, both upstream and downstream of farming. Concentration has increased still further since 1994, exacerbated by the entry of multinational seed, fertilizer and agrochemical companies (Bernstein, 2013, pp. 29–30).

Contrary to the predictions of the World Bank, whose proposals for Rural Restructuring informed the deregulation plan of the 1990s, these policy reforms neither lowered barriers to entry for small farmers, nor lowered the price of food. Food price inflation over the past 15 years has far exceeded general inflation, not only due to widespread collusion in key sectors – notably fertilizer, seed, poultry, milling, bread, and retail – but due to the concentrated structure of the food system itself, and its industrial farming models premised on input- and capital-intensive production. The negative social outcomes of this economic structure have therefore become apparent and widely recognized.

State responses do not confront agrarian capital in its various forms, but accommodate and even support it – despite the creation of a Competition Commission that investigates anti-competitive behaviour within a limited mandate. The ruling African National Congress has consistently supported large-scale commercial farming, confronting only those market 'distortions' brought about through cartel behaviour, rather than the fundamentals of the structure of the economy. However, given a growing crisis of social reproduction of 'fragmented classes of labour', only partially defused by social grants payments, capital-friendly policies such as these are increasingly coming under fire. In short, the contradictions of South Africa's capitalist agro-food system are increasingly evident (Greenberg, 2017).

4. The Increasing regional footprint of South African agrarian capital

South African agrarian capital, in the form of both large-scale commercial farming operations and agribusiness firms that are active up- and down-stream of farming itself, is expanding across Africa (Boche & Anseeuw, 2013; Hall, 2011; Hall, 2012). Here we examine why and how this is taking place, and with what success. Table 1 lists companies located in different nodes of agricultural value chains, which are active in different African countries, many of which we discuss further below.

Reforms to lower barriers to intra-regional trade have been key to the expansion of South African companies, giving them a competitive edge over some of their multinational competitors. The South African Development Community (SADC) Free Trade Area of 2008 set out a phased reduction in tariffs on intra-regional trade, facilitating more investment and cross-border sales of agricultural produce and foodstuffs. Expanded market access was provided from 2015 by the extension to a tripartite free trade area, across the regional blocks of SADC, the East African Community and the Common Market for East and Southern Africa (COMESA).

4.1. Financialization and farmland funds

Along with deregulation of South African agriculture has come financialization, a feature currently being exported elsewhere in the region. Reflecting global trends (Fairbairn, 2015) financialization has

Table 1. South African agro-food corporation expansion into Africa and beyond.

Country	Node of value chain	Corporation
Africa		
Angola	Fertiliser	Omnia
	Logistics	Barloworld
	Wholesale and retail	Shoprite, Spar
Botswana	Primary production	Country Bird
	Agri services	Unitrans
	Logistics	Barloworld, Bidvest, Imperial, RCL Foods
	Food manufacturing	AVI, Pioneer Foods
	Wholesale and retail	Massmart, Pick n Pay, Shoprite, Spar, Woolworths
Cameroon	Food manufacturing	Tiger Brands
Cape Verde	Logistics	Barloworld
DRC	Logistics	Barloworld
	Wholesale and retail	Shoprite
Ethiopia	Food manufacturing	Tiger Brands
Ghana	Agri services	AFGRI
	Logistics	Bidvest, Imperial
	Wholesale and retail	Massmart, Shoprite, Woolworths
Kenya	Food manufacturing	Tiger Brands
	Logistics	Bidvest, Imperial
	Wholesale and retail	Woolworths
Lesotho	Agri services	Unitrans
	Logistics	Barloworld, Bidvest, Imperial
	Fertiliser	Omnia
	Food manufacturing	Premier Foods
	Wholesale and retail	Massmart, Pick n Pay, Shoprite, Spar, Woolworths
Madagascar	Agri services	Unitrans
	Wholesale and retail	Shoprite
Malawi	Agri services	Unitrans
	Fertiliser	Foskor
	Logistics	Barloworld, Bidvest, Imperial
	Wholesale and retail	Massmart, Shoprite
	Manufacturing	Illovo
	Primary production	Illovo
Mauritius	Fertiliser	Omnia
	Logistics	Bidvest
	Wholesale and retail	Shoprite, Woolworths
Mozambique	Grain/feed milling	Astral
	Primary production	Country Bird, Illovo, Tongaat Hulett
	Agri services	Unitrans
	Fertiliser	Omnia
	Food manufacturing	Illovo, Premier Foods, Tongaat Hulett
	Logistics	Barloworld, Bidvest
	Wholesale and retail	Massmart, Shoprite, Spar, Woolworths
Namibia	Retail input	Zeder
	Primary production	Country Bird
	Agri services	Unitrans
	Food manufacturing	AVI, Pioneer Foods
	Logistics	Barloworld, Bidvest, Imperial
	Wholesale and retail	Massmart, Pick n Pay, Shoprite, Spar, Woolworths
Nigeria	Food manufacturing	Tiger Brands
	Logistics	Imperial
	Wholesale and retail	Massmart, Shoprite
Sao Tome and Principe	Logistics	Barloworld
Seychelles	Logistics	Bidvest
Swaziland	Primary production	Astral, Illovo, Tongaat Hulett, TSB
	Agri services	Unitrans,
	Fertiliser	Foskor, Omnia
	Grain/feed milling	Premier Foods
	Food manufacturing	Illovo, Premier Foods, TSB
	Logistics	Barloworld, Bidvest, Imperial
	Wholesale and retail	Massmart, Pick n Pay, Shoprite, Spar, Woolworths

(Continued)

Table 1. Continued.

Country	Node of value chain	Corporation
Tanzania	Agri services	Unitrans
	Logistics	Bidvest, Imperial
	Manufacturing	Illovo
	Primary production	Illovo
	Wholesale and retail	Massmart, Woolworths
Uganda	Grain/feed milling	Quantum
	Logistics	Bidvest
	Wholesale and retail	Massmart, Shoprite, Woolworths
Zambia	Primary production	Country Bird, Illovo, Zeder
	Grain/feed milling	Astral, Country Bird, Quantum, Zeder
	Agri services	Afgri, Unitrans
	Fertiliser	Foskor, Omnia
	Food manufacturing	AVI, Illovo, RCL Foods
	Logistics	Barloworld, Bidvest, Imperial
	Wholesale and retail	Massmart, Pick n Pay, Shoprite, Woolworths
Zimbabwe	Primary production	Country Bird, Tongaat Hulett
	Agri services	Afgri
	Fertiliser	Foskor, Omnia
	Food manufacturing	Tiger Brands, Tongaat Hulett
	Logistics	Barloworld, Bidvest, Imperial
	Wholesale and retail	Pick n Pay, Spar
Outside Africa		
Middle East and Australasia	Agri services	Afgri
	Fertiliser	Omnia
	Logistics	Barloworld, Bidvest
Americas	Fertiliser	Omnia
	Food manufacturing	Tiger Brands
	Logistics	Barloworld, Bidvest
Europe and UK	Food manufacturing	Pioneer Foods
	Logistics	Barloworld, Bidvest, Imperial
	Wholesale and retail	Spar

Source: Greenberg (personal communication).

involved the emergence of South African-based agricultural investment funds (Boche & Anseeuw, 2013). Animating this is a growing cast of actors through whom transnational private capital is being brought into Africa's agriculture, ranging from pension funds, hedge funds, sovereign wealth funds, banking institutions and agribusinesses and private equity funds. Among their financial instruments are 'farmland funds' offering share portfolios – essentially creating a new asset class. An influential European report on the 'Vultures of Land Grabbing' characterized such funds as 'not only [having] a speculative business model, but also represent[ing] a conveyor belt for shareholder capitalism from the financial to the real economy' (Merian Research and CRBM, 2010, p. 3).

Several such funds with regional and even pan-continental ambitions were established in South Africa from 2008 onwards, among them being Emergent Asset Management Ltd, a UK/SA management firm emerging from defence and high-tech industries in the US and now specializing in farmland investments in Africa. In the midst of global recession, its African Agricultural Investment Fund, established in 2008, aims to grow to €3 billion and promised its large institutional investors 30 percent annual returns. Partnering with Grainvest, one of the top 5 companies on South Africa's SAFEX, it formed operating company Emvest Agricultural Corporation, providing a vehicle for South African, UK and other investors to diversify their investments into African agriculture in Angola, Botswana, Democratic Republic of Congo, Kenya, Lesotho, Madagascar, Malawi, Mauritius, Mozambique, Namibia, South Africa, Swaziland, Tanzania, Zambia, and Zimbabwe (McNellis, 2009, p. 13). Susan Payne, formerly of Goldman Sachs, initiated Emergent and formed the South African connection, punting African farmland as the most promising new investment frontier for institutional investors and wealthy individuals, arguing that 'because of its series of microclimates, its highlands, its agricultural diversity and good logistics, South Africa and sub-Saharan Africa can deliver an enormous amount of food' (Payne, cited in McNellis, 2009, p. 13).

More 'homegrown' among the South African farmland funds is United Fruit Farmers and Agri Asset Management, part of investment and insurance company Old Mutual's African Agricultural Fund. This vehicle aims to take 'advantage of Africa's enormous untapped agricultural potential' through twin funds – one for internal acquisitions through Futuregrowth Agri-Fund (SA) and the other being the African Agricultural Fund, enabling South African investors to channel investments both internally in the domestic market and externally in the region (UFF, 2015). Alongside these are several other funds, with a range of internal and regional foci including, among others, Phatisa's African Agricultural Fund which by 2016 held equity in the region of half a billion US dollars (Anseeuw et al., 2012).

Alongside these new financial actors are the more traditional forms of finance, both public and private. Among these are South African banks Standard Bank[1] and ABSA, themselves transnationalized (Hall, 2011). From the agribusiness sector is Afgri, a privatized state-established farming cooperative, formerly Oos-Transvaal Beperk (OTK), which has, by absorbing other former state cooperatives, reinvented itself as a leading agribusiness in inputs, including through a continent-wide license to sell John Deere tractors, and is the main source of finance and logistics for several farmland investments, including the South African farming venture in Congo (Hall et al., 2015). It has, in turn, been absorbed by the Canadian company Fairfax which bought a controlling share in Afgri, despite protestations by the African Farmers' Association of South Africa which objected that this once-parastatal was being sold off to a multinational rather than into black South African hands.

Changes to financial regulation would incentivize investment from and via South Africa by creating 'simpler rules' to reduce the time and costs of doing business in Africa. By allowing holding companies exempt from the South African Reserve Bank's exchange controls to be created by companies listed on the Johannesburg Stock Exchange, as regional investment vehicles which would not be regarded as resident for exchange-control purposes. 'Similar measures [would] apply to foreign companies wanting to invest in African countries using South Africa as their regional headquarters … as part of the Gateway to Africa reforms … , including BRIC countries' (Gordhan, 2013, p. 12–13).

Less significant by far is the direct role of state finance through development finance institutions (DFIs), primarily the Industrial Development Corporation (IDC) with its massive infrastructure projects in road, rail, energy, and mining, and the Development Bank of Southern Africa, which has significantly increased its financing of regional projects in transport, energy, mining, ICT, health, financial services and manufacturing, primarily to Zambia, and Mozambique (Govender, 2013, pp. 10, 16). Meanwhile, the Agricultural Business Chamber has spearheaded studies on agricultural market opportunities in Sudan, Uganda, Ethiopia, Egypt, and Kenya, providing advice to South African agribusinesses in support of their expansion plans (ABC, 2012). Clearly, faced with economic slowdown at home, capital is on the move out of South Africa but also globalized capital is on the move across Africa, in part via South Africa. This suggests that a complex interplay of South African conditions, global trends and conditions in host states are combining to see agriculture become increasingly financialized, as new institutional actors cash in on African farmland as a new investment frontier.

4.2. Inputs: seed, fertilizer and pesticides

From an apartheid past where major agribusiness companies were state-owned, major seed, fertilizer and pesticide companies and cooperatives have been privatized, and several have reinvented themselves as regional players. The influence of multinational and South African input supply industries in Africa's agro-food system has grown alongside their consolidation in seed, pesticide and fertilizer markets. One way in which this is happening is via multinational corporations buying up or into South African corporations and these in turn expanding in the region. Examples include Pannar Seeds (now largely owned by Du Pont), together with Monsanto and Pioneer Seeds, all but monopolizing the local market for maize, sorghum and wheat seed.

One of South Africa's seed giants, Pannar, was acquired by Du Pont's Pioneer Hi-Bred in 2013, effectively consolidating the domestic seed market in the hands of just two companies: Pioneer and Monsanto. This was initially prohibited by South Africa's Competition Commission on the grounds that it would give Pioneer an anti-competitive advantage, a ruling that was confirmed by the Competition Tribunal but later overturned by the Competition Appeal Court (ACB, 2015, p. 25). The merged company and Monsanto together now hold 90% of the South African seed market for maize, wheat, and sorghum (Bernstein, 2013, 30). The merger, according to Pannar CEO Deon van Rooyen, would give the South African company access to the US company's technology and research, and in turn offer it access to South African maize germplasm and a base from which Pioneer could 'reach farmers it is not currently serving, such as those in some of the small-scale farming regions locally and elsewhere in Africa' (cited in Coleman, 2012). Among these would be Pannar's existing clients in other countries, where it holds significant market share in Zimbabwe (18%), Zambia (15%), and Tanzania (15%) (ACB, 2015, p. 25).

As well as seed, in other sectors too, South African and multinational companies are leading the way in transforming the pesticide market, which is now dominated by global companies – Monsanto, Pioneer, Syngenta, and a few others, their entry into African agriculture facilitated in part by the G8's New Alliance for Food Security and Nutrition. Further mega-mergers have been proposed, including ChemChina and Syngenta, DuPoint and Dow, and Monsanto and Bayer (Who Owns Whom, 2017a), with the latter being particularly contentious in South Africa where state approval for the merger is still pending. Despite there being some regulatory brakes, then, the pesticide market, like seed, is increasingly concentrated and multinationalized. Though Syngenta and Monsanto have direct market links into African countries – not necessarily using South Africa as a launching pad – there are many grain-trading joint ventures in the region that pair up South African companies with these multinationals. For instance, multinationals Cargill and Dreyfus are dominant grain traders in South Africa, handling 70% of maize trade, and from there also expanding also elsewhere in the region (Greenberg, 2017).

South Africa's main chemical fertilizer companies, Sasol and Omnia, now operate as multinationals, and were found guilty – and fined – by South Africa's Competition Commission in 2009 for cartel behaviour, together with the Norwegian-based Yara International. These companies, along with a small number of other multinationals, are the primary beneficiaries of subsidies under the rubric of the 'Green Revolution in Africa', funded by USAID and other donor agencies, including the Bill and Melinda Gates Foundation (ACB, 2014). The G8's New Alliance on Food Security and Nutrition and the US's Grow Africa initiative both involve policy concessions from priority countries to open their markets to such companies, as part of their 'country cooperation frameworks' which are required to unlock donor funds. South Africa's Omnia, now in 10 countries in Africa, has expanded in recent years but is dwarfed in this market already dominated by several global (and a few African) firms, mostly notably Yara, whose profits rose threefold on the back of global price increases between 2009 and 2012 (ACB, 2014, p. 31). So even as South African input industries expand, they are, at times, in competition with larger actors from the global North, but at other times have agglomerated via mergers and acquisitions. By and large, we see the South African seed, pesticide, and fertilizer companies being consolidated as well as being acquired by multinationals, as they expand regionally.

4.3. Land deals

South African farmers and companies have been identified as among the 'land grabbers' in Africa, with land deals being premised on the expansion of South African industrial farming systems to new sites on the continent – but this story is complex and has changed over time (Hall, 2011; Hall et al., 2015). South African (white) farmers and farming companies have been on the move, securing substantial land deals across several countries – Mozambique, Zambia, and Congo being prime among them – while complaining of low profitability, high costs, and political threats to farmers at home. Africa has been at the centre of the 'global land grabbing' phenomenon from about 2008 onwards, arising from the food price crisis of 2007/2008, the financial crisis of 2008 and subsequent global recession, and fuel price spikes around the same time. While the reasons for the preponderance of large-scale corporate land deals in Africa are debated (Scoones et al., 2014), what is clear is that South African companies are significant among the range of actors involved. Deals are being struck typically between South African companies and foreign governments, sometimes with local business partners. They are not exclusively for food production or even for agriculture, as South African companies are now engaged in forestry deals in Mozambique and Ghana, in farming projects in Congo, Mozambique, Swaziland, Zambia, Zimbabwe, and Nigeria, and in tourism (wildlife safaris and ecotourism) in Mozambique, Tanzania, and Uganda (Matrix, 2015). These sit alongside general banking, financial services, telephony, construction, and information technology investments by South African firms.

Facilitating many of these deals is the commercial farmer association, Agri South Africa (AgriSA), which in 2013 created 'AgriAllAfrica (AaA) as an external agricultural investment facilitation platform to enable South African farmers to invest more easily in farmland and agriculture elsewhere on the continent:

> South African farmers have started to spread their wings considerably wider than the traditional South [ern] (sic) African Development Community (SADC) … The international focus on agriculture's potential in Africa has further intensified over the past year, with an increase in investments in various high-potential agricultural countries. (AgriSA Africa Policy Committee, 2014, p. 35)

From the gung-ho plans to secure land concessions in 22 African countries in the 2011–2012 period, AgriSA has, after disappointing results of its farmer groups 'AgriSAMoz' in Mozambique and 'Congo Agriculture' in Congo, withdrawn to more modest aims of consolidating its members' operations through acquiring 'priority status' as agricultural investors in these countries, and engaging in talks with host governments to provide further protection and support. Neoliberal visions and post-colonial ambitions have foundered in practice, and the 'land grab bubble' has burst. Meanwhile, AgriSA has been pursuing new opportunities for land concessions in Ethiopia and Nigeria, while continuing to monitor conditions in countries where initial talks have been held, including Botswana, Swaziland, Tanzania, Angola, DRC, Uganda, Rwanda, Sudan, South Sudan, Eritrea, Egypt, Chad, Ghana, Gabon, and Sierra Leone (AgriSA Africa Policy Committee, 2014, p. 39). Meanwhile, at home, it is lobbying the Ministry of Trade and Industry for support in securing further sites in the face of South Africa's discontinuation of bilateral investment treaties.

AgriSA's 'Africa Policy Committee' reports that 'The international focus on agriculture's potential in Africa has further intensified over the past year, with an increase in investments in various high-potential agricultural countries' (AgriSA Africa Policy Committee, 2014, p. 35). In 2013, it created an investment platform named 'AgriAllAfrica' (AaA) to facilitate South African farmers' deals in farmland and agriculture elsewhere on the continent, and brought state representatives from other African states on visits to South Africa to broker deals. Meanwhile, commercial farmers are forging stronger relations with regional farmer bodies, notably the Southern African Commercial Agricultural Union (SACAU) and the new continent-wide alliance of regional farmer organizations, the Pan African Farmers' Organization (PAFO). Having already headed SACAU, AgriSA's chief land deal negotiator, Theo de Jager, now heads both organizations, having been elected as president of PAFO in 2014, and has used his influence in the region and across the continent to promote commercial agriculture and increased uptake of technology, and regional integration through intra-African investment and trade (PAFO, 2014). On his election as president of PAFO last year, he set out priorities for African agriculture, including 'a change of mindset from fighting poverty through agriculture, to wealth creation', and a need for Africa to take ownership of opportunities on the continent primarily through intra-Africa trade (PAFO, 2014). This reflects a distinctive neoliberal development ideology being advanced by South African farmers.

One of the main 'success stories' of South African agribusiness on the continent is that of sugar, notably the SA sugar giants, Illovo, and Tongaat Hulett, each with operations in six countries in the region and, to a lesser but growing degree, also TSB, now active in three countries. The success of these sugar companies builds in large part on the export of a model developed and honed over decades in South Africa, of nucleus estates supplemented by (indeed, often largely dependent on) contracted outgrowers (Dubb, 2016). The adaptation of this model of contract farming, in different ways from Tanzania to Malawi to Zambia to Mozambique, shows how these large companies have on the one hand reproduced production systems and labour regimes, and the social relations that underpin them, while also varying their modalities. At the same time that they have expanded their regional footprint, their ownership structures have also changed, with the most significant regional player, Illovo, now being 100% owned by Associated British Foods. While retaining its South African base of operations, like many others, it is no longer a South African company. That sugar is at the 'frontier' of expanding industrial farming in the region resonates strongly with its important historical role as a frontier crop in the global expansion of capitalism (Moore, 2015).

While land deals may constitute processes of 'accumulation by dispossession' (Harvey, 2003), other factors are also changing landholdings. As Jayne, Chamberlin, and Headey (2014) have shown, endogenous concentration in landholdings is underway in several African countries, driven by local and national elites, and this dynamic, including market transactions and local elite land grabs, possibly overshadows transnational corporate 'land grabs' as a driver of concentration. This suggests that domestic capital within African countries is moving up and downstream through food value chains. These are highly dynamic contexts into which South African capital, and capital routed via South Africa, is expanding.

4.4. Food processing, manufacture, logistics, and distribution

South African companies are also exporting their food processing, manufacture, logistics, and distribution operations, as domestic demand stagnates. Building on the expansion of sugar are the four South African food giants – Tiger Brands, Pioneer Foods, Premier Foods, and RCL Foods[2] (formerly Rainbow Chicken Limited) – which together dominate processing and manufacture in South Africa. These companies are at the epicentre of South Africa's tightly controlled food value chains. In the 'bread cartel' scandal of 2009–2016, the Competition Commission found three of these – Tiger Brands, Pioneer and Premier – to have colluded to fix prices by using their collective market power which spans milling and baking subsidiaries (Cock, 2009).

Over the past decade, these companies have regionalized, prompted by dulled growth prospects, regulatory constraints and competition at home and by improved prospects in regional markets. For instance, faced with domestic competition from cheap Brazilian, EU, and USA chicken imports – the latter being the result of a bilateral trade deal between South Africa and the USA in 2016 under the aegis of the Africa Growth and Opportunity Act (AGOA) – RCL Foods was joined by the poultry association in lobbying government to conclude government-to-government deals with other countries in the region to facilitate the export of South African chicken into neighbouring markets (Pitso, 2016). As well as rising market competition at home, another push factor driving poultry players into regional markets is the impending introduction of more stringent regulations on the maximum level of 'brining'[3] from 30% to 15% – regulations not applicable in neighbouring countries (Pitso, 2016).

As a result, the South African poultry industry is rapidly regionalizing. RCL acquired a 49% in Zam Chick in 2013 for $14.5 million, the year it also acquired FoodCorp, South Africa's third biggest food producer, for $113 million and changed its name from Rainbow Chicken Limited to RCL Foods Limited (IOL, 2013). In 2015, together with Zambeef, the majority owner of Zam Chick (Zambia Chicken), it made a $4 million investment in Zam Hatch, thereby extending its involvement in Zambia's poultry value chain from a feedmill to laying farms, a hatchery and rearing farms (RCL Foods, 2016b) – thereby achieving a rapid degree of vertical integration. RCL discourses reflect this turn to an outward-facing business strategy:

> RCL Foods has implemented a business model to grow the company and ultimately supply high quality, nutritious and affordable food to the entire African continent … we have begun to actively expand our reach beyond South African borders by continually seeking out and negotiating with prospective new business partners across Africa. Through our joint venture partnerships with leading food supply companies in the African market, we will branch out into new and exciting avenues which will benefit the Group as well as our African associates. (RCL Foods, 2016a)

The experience of the dairy industry offers some cautionary tales. Clover, one of South Africa's largest dairy producers in the domestic market, withdrew in 2013 from a supply agreement with Danone, the Italian company with a global footprint, and now faces competition from it elsewhere on the continent where Danone has 19 factories and a workforce of 10,000 (Peacock, 2016). Clover's attempts to break into the Nigerian and Angolan dairy markets faltered, as 'logistical and supply-chain challenges thwarted attempts to build its brands' in those countries (Peacock, 2016). Its competitor, Danone, with deeper pockets for more substantial investment, has in contrast bought controlling stakes in major dairy companies in both Nigeria and Kenya, tapping into their existing and extensive distribution systems, which include 25,000 'pushcart bicycle sellers on the streets' – something which the South African company Clover, by 'going it alone', could not do (Peacock, 2016). All this suggests some binding constraints faced by South African companies attempting to expand in isolation into the agro-food systems of other countries on the continent.

Tiger Brands, the biggest food manufacturer in South Africa, is operational in 22 countries in Africa, with a focus on maize value chains. Facing declining profits at home, Tiger Brands has in the past decade embarked on aggressive acquisitions in other African countries, acquiring Nigerian biscuit manufacturer Deli Foods in 2013, as well as a 51% stake in the Ethiopian food and beverage East African Group, and 49% of the food and beverage operations of UAC of Nigeria Plc (Africa Business Journal, 2014). But the direct acquisition of manufacturing businesses has not fared well, and Tiger Brands has withdrawn from this mode of operation, shifting instead to investing in distribution networks for its South African processed foods. For example, in 2016 it sold its biggest investment in Nigeria after an R1 billion loss, which also led to the axing of its CEO (Goko, 2016). Meanwhile, its arch-rival at home, Pioneer Foods, has also been expanding its Africa operations, having acquired a majority stake in Food Concepts PLC, its Nigerian rival in the fast food and bakery sector, and focusing on further expansion of operations in Angola, Kenya, Ethiopia, Tanzania, and Ghana. Pioneer has extended its footprint through vertical integration combined with regional expansion, into grains, animal feed, poultry, and beverages – until recently through a licensing agreement with PepsiCo (Pioneer Foods, 2014).

A feature of regional expansion by South African companies is that they *co-expand*. There are several examples of the regionalization of existing business partnerships. For instance, Unitrans provides transport and logistics services for Illovo sugar estates and to its outgrowers in South Africa, and now also at its operations in Malawi, Mozambique, and Tanzania (Smalley, Sulle, & Malale, 2014). Unitrans now operates in 10 countries in Southern Africa, providing logistics, leasing of on-farm machinery, storage, transport, and supply-chain management services, often in partnership with South African-based companies, including Tiger Brands and RCL Foods (Unitrans, 2017).[4] This shows how South African companies moving into the region tend to pull their value chains with them, often in the context of long-established commercial relationships forged inside South Africa, and sometimes in the absence of equivalent service industries in host countries. Unitrans is a wholly owned subsidiary of KAP, Industrial, established in 2003 as 'a diversified industrial business focused on growth in African markets' and which acquired Unitrans sometime after 2004.[5] In this way, Unitrans, first established in 1962, became part of a larger conglomerate of companies focused on complementary service provision to manage regional logistics and supply chains. All this shows how business networks shape regional expansion, and how South African companies with histories of partnership rely on one another, effectively pulling their value chains with them as they venture into new territory.

4.5. Supermarkets and fast food

At the retail end of the value chain, South African supermarkets and fast food chains are rapidly expanding their reach, aiming to cash in on both the growing 'middle class' market in African cities and the growing low-end market for cheap manufactured foods. Four giant South African supermarket chains – Shoprite, Pick n Pay, Spar and Woolworths – have all developed a regional imprint over the past two decades. Shoprite, 'arguably the most successful supermarket chain in sub-Saharan Africa', is at the forefront, with 320 supermarkets (both corporate and franchise) in 14 African countries, and with rapid expansion both within and beyond these countries (Harding, 2011). Sensitive to criticisms of South African companies bypassing local producers, Shoprite has established programmes to assist producers of fresh fruit and vegetables to meet its quantity and quality requirements, and so to indigenize its procurement practices in its other African operations (Shoprite Holdings Limited, 2016). Yet, it was still importing 98% of its retailed fresh fruit and vegetables from South Africa; the same proportion of 98% of manufactured food sold in Namibia is imported from South Africa (Emongor, 2008). By 2014, Shoprite CEO Whitey Basson's message to his shareholders was summed up as: 'Don't bother looking at South Africa — Africa is where the really profitable action is' (Shevel, 2014). Citing the growing middle classes in African capital cities, and their desire for big brands and access to a greater variety of manufactured and processed foods, he cautioned his own investors to 'ignore the potential of Africa's shoppers at their peril' (Shevel, 2014). Clustering its food retail (Shoprite, Checkers, Checkers Hyper, USave, OK Foods, and OK Grocer) together with fast food (Hungry Lion), liquor (OK Liquor, Friendly Liquor) and its furniture (OK Furniture, OK Furniture Dreams, and House & Home) and pharmacy retail brands (Shoprite MediRite) (Shoprite Holdings Limited, 2016).

Far behind Shoprite is Pick n Pay, with its Zimbabwean subsidiary TM, which has been expanding in Zambia and Ghana, preparing to move into Nigeria, while closing operations in Mozambique and Mauritius. Food Lovers' Market has also expanded from South Africa northwards into Zimbabwe. Less significant but still present in food retail across several African countries are Spar and Woolworths. High-end chain Woolworths opened three stores in Nigeria in 2012, closing all by 2014 (Goko, 2016). By 2016, though, Pick n Pay was aiming to expand into Nigeria, this time in a 51–49% partnership with a listed Nigerian company, and initially opening a range of 10 large and small format stores (Goko, 2016).

Of more concern to Shoprite than its South African competitors is the Kenyan supermarket chain Nakumatt, and Walmart (Harding, 2011) and, to a lesser degree perhaps, Botswana's own supermarket chain Choppies, now also in South Africa and elsewhere. Kenya's Nakumatt supermarket chain now has 35 retail outlets, spanning Kenya, Uganda, Rwanda, and Tanzania, taking over Shoprite's flagship Dar Es Salaam outlet in 2014, and aiming to expand to Burundi 'to ensure that we fully cover East Africa before setting off on the Nakumatt 2.0 journey, which involves registering a Pan-African presence' with the next steps being supermarket expansion in Nigeria, DRC, South Sudan, Malawi, and Botswana, said managing director of Nakumatt Holdings, Atul Shah (Harding, 2011).

Also on the move is a fifth supermarket giant in South Africa, Massmart, which had existing outlets in 11 countries in Africa (Massmart, 2012) prior to US supermarket giant Walmart's contentious purchase of it in 2010. This acquisition was contested by the national trade union federation, the Congress of South African Trade Unions but approved by the Competition Commission. As a result, Walmart, via Massmart, has expanded its African footprint through its new holdings in the food wholesale and retail sectors through its subsidiaries Game and CBW, both of which had existing outlets in several countries. Since this purchase, Massmart has moved into more direct competition with other South African food retailers in West Africa, including Shoprite which already has 16 stores in Nigeria and which, by diversifying its supply chains, now procures 76% of food items from local suppliers and farmers ('Diyan, 2016). Massmart is competing for the same market, while also opening smaller grocery stores aimed at Nigeria's vast low-income market. Rather than eyeing competition from local retailers, or investors from elsewhere, Massmart identifies other South African retailers as its primary competitors: 'Shoprite was first and we're second but with the power of Walmart we hope to overtake them' (Grant Pattison, CEO Massmart, cited in Ventures Africa, 2014).

As South African, Kenyan and other supermarkets expand, first into national capitals and then into smaller towns, procurement practices and quality and quantity requirements limit access by local smallholders into formal food retail chains (Weatherspoon, Neven, Katjiuongua, Fotsin, & Reardon, 2004). Such negative impacts reflect the broader contradictions between corporate food retail and more socially embedded food trade networks, and the tendency of the former to displace the latter (Weatherspoon & Reardon, 2003). This is another way in which South African agro-food capital is exporting its contradictions and tendency to exclude small-scale farmers and rely on large-scale procurement of inputs, large-scale farming, large-scale processing and manufacture, and large-scale retail.)

5. Understanding the conditions for failure

Despite the far-reaching expansion, success for South Africa as a (potential) regional hegemon in Africa's agro-food system is far from assured. Rather, we see diverse cases of success and failure by individual companies. This is why we find it instructive to look not only at where South African capital has succeeded – in cases lauded on the popular South African business website 'How we made it in Africa' – but also to understand the reasons for its failures.

Several significant examples of failure suggest a range of reasons, including competition from domestic agro-food capital; competition from regional actors from other middle-income countries; and supply-chain difficulties including input and output markets. Among the 'failures' perhaps the most notable is Tiger Brands' purchase in 2012 of a majority (65.7%) shareholding in Nigeria's Dangote Flour Mill, an investment which lost nearly a quarter of its market value in the year 2013–2014 before Tiger Brands pulled out. Second, also in Nigeria, is the demise of the Woolworths retail venture, which foundered on supply-chain problems and high pricing of its clothing lines, unlike its competitor Shoprite which, through South African-Nigerian joint ventures, managed to get anchor positions in 10 new malls at the time of the demise of Woolworths Nigeria (Douglas, 2014). Third, Pioneer, too, divested its unprofitable subsidiary Quantum Foods, which owns the milling and food manufacture companies Bokomo Uganda and Bokomo Zambia. Fourth, after an initial expansion into Tanzania, all three of Shoprite's stores in the country, in Dar es Salaam and Arusha, were bought out by Nakumatt for $45.5million in 2014 (Ciuri, 2014; Who Owns Whom, 2017b). Fifth, several farmland investments have dwindled from ambitious initial plans to a modest scale of farming. For instance, in Congo, a decline from a plan of 10 million hectares, to agreement on 200,000 ha, to initial allocation of 80,000 ha, only a fraction of which has been cultivated, and with massive attrition of initial investors, with most returning home (Hall et al., 2015), echoing the experience of white Zimbabwean farmers who moved into Mozambique after fast-track land reform (Hammar, 2010).

We hypothesize that one reason for failures such as these is the fact that not all elements of the system – which have been combined to create South Africa's corporate agro-food system – are present in new destination markets. The export of some elements of this model into countries where states have neither the capacity nor the willingness to create a capitalist farming class through regulation and subsidy in the way that the apartheid government did, does not always gain traction, and in practice has foundered on numerous occasions. Land deals for primary production depend on the provision of both infrastructure and political support to facilitate the movement of big capital through the agro-food system, as the South African state provided for several decades. The absence of such conditions makes corporate investments vulnerable to conditions that differ considerably from those that enabled accumulation at home. Capital is thus vulnerable to a degree, suffering setbacks and losses in the face of an absence of the conditions that enabled the development of a capitalist agro-food system in South Africa in the twentieth century.

6. Conclusions

This article explores the expanding role of forms of agrarian capital based in one of the BRICS countries, South Africa, in agro-food systems in Africa more widely. We have noted the ways in which the structure of the South African economy has both conditioned and precipitated this process of regionalization. We have argued that there are a number of significant 'push' factors, including stagnating domestic demand in the context of massive, structural unemployment, with capital driven to seek new markets for products. Against this backdrop, we argue that the end of apartheid and the increased degree of integration into the global economy that this entailed came at a good time for South African capital. Here we concur with Bernstein (2013, p. 23) who notes 'the importance of the removal of the limits on international mobility of capital and commodities imposed by the apartheid regime' and the ways in which these policies, and their timing, have enabled South African agribusinesses to reposition themselves in the era of deregulation and liberalization.

However, capital moving from and via South Africa into other Africa countries has also encountered substantial obstacles to entry, and been challenged by severe competition from other companies in its destination markets. These companies are located both in and from other middle-income countries with expanding agro-food industries of their own. While a 'first-mover advantage' may have buoyed South African corporate investors in the late 1990s and early 2000s, since then their expansion has been more chequered, and seen several notable failures as well as some successes. The ambitions of giant South African agribusinesses to become regional hegemons are thus often thwarted.

Apartheid policies from the 1940s onwards established the conditions for the emergence of a particular form of agrarian capital, corporate agribusiness, in which an authoritarian state was able to enforce low levels of subsistence while simultaneously increasing the productivity and output of capitalist agriculture through subsidies made possible by funds from a lucrative mining sector. In other African countries, given their different histories and circumstances, these conditions rarely exist.

South African corporate investment in countries across the continent – in the food system, but also in mining, telecommunications, finance, construction, transport, and logistics – often involves companies retaining their South African base. But, as the example of major brewer SABMiller illustrates, many companies are 'transnationalizing', rebranding themselves for an African market, while providing a route through which global capital can partner with South African capital in its expansion strategies. As the examples of the Chinese stake in Standard Bank and the purchase of Illovo Sugar by Associated British Food plc show, the notion of 'South African capital' is becoming increasingly moot.

Understanding the links between internal and regional agro-food transformations requires that we examine both 'push' and 'pull' factors conditioning the behaviour of capital, as well as the impacts of the entry of South African companies. These are the core of an agenda for future research. The key questions can be posed as follows: first, which actors in agro-food value chains are the key drivers of regional expansion – for instance, is this primarily a production-, processing-, or retail-led dynamic? Second, what are the conditions that make exporting elements of the South African agro-food system feasible or unfeasible, and influence the outcomes of such investments – in other words, what explains the variable track record, and outcomes of 'success' and 'failure', and for whom? Third, how do 'host states' position themselves in soliciting inward investment while also aiming to support local ownership of core farming, processing, and retail companies, and how does the character of the local political economy shape the terms on which outsiders – South Africans and others – are allowed to enter into Africa's agro-food system? Fourth, where is the South African state in all this? How is the state positioned with respect to agribusiness and related capital, South African and multinational, in an era of deregulated commodity and financial markets?

In this article we have argued that South African agrarian capital's engagement elsewhere on the continent exhibits a degree of path-dependency, reflecting its domestic accumulation path and its socially and politically contradictory focus on benefiting a narrow (racially defined) elite. However, the ways in which this plays out in practice are highly contingent, as capital encounters different conditions and new competitors. What is under way is the export of elements of South Africa's agro-food system into countries where states lack either the capacity or the willingness (or both) to create a small, elite class of capitalist farmers, through regulation and subsidy, in the way that the apartheid government did. It is for this reason that, despite the narrative of 'Africa rising' and the allure of growing markets, South African capital suffers not from only competition but also severe setbacks, often due to the starkly different conditions into which elements of South Africa's agro-food system are inserted. Export of this system is thus built on shaky foundations.

Notes

1. Standard Bank now also operates in Angola, Botswana, Congo, Ghana, Kenya, Lesotho, Malawi, Mauritius, Mozambique, Namibia, Nigeria, Swaziland, Tanzania, Uganda, Zambia, and Zimbabwe. But through acquisitions it has also extended beyond Africa to the Americas (Brazil, Argentina, and the USA) and to China, Hong Kong, Isle of Man, Japan, Jersey, Singapore, Taiwan, Turkey, United Arab Emirates, and United Kingdom.
2. RCL Foods, formerly Rainbow Chickens Limited, changed its name in 2013 to reflect its wider spectrum of brands, and is majority owned by Remgrow.
3. Brining is the injection of brine – salty water – into the chicken, ostensibly to retain succulence and improve flavour, thereby increasing weight without commensurate nutritional benefit.
4. http://www.unitrans.co.za/customers
5. http://www.unitrans.co.za/about http://www.kap.co.za/about/our-history/ http://www.kap.co.za/ – but contradicted by this: http://www.saflii.org/za/cases/ZACT/2005/9.pdf about Steinhoff and forestry expansion.

Acknowledgements

We thank the journal's peer reviewers for their constructive and insightful reviews that helped us to improve the quality of this article.

Disclosure statement

No potential conflict of interest was reported by the authors.

Funding

This work is based on the research supported by the South African Research Chairs Initiative of the Department of Science and Technology and National Research Foundation of South Africa (Grant No. 71187).

References

ABC (Agricultural Business Chamber). (2012). 'Trade intelligence: Country profiles.' Retrieved from http://www.agbiz.co.za/TradeIntelligence/CountryProfiles/tabid/466/Default.aspx

ACB (African Centre for Biodiversity). (2015). *The expansion of the commercial seed sector in sub-Saharan Africa: Major players, key issues and trends*. Johannesburg: ACB. Retrieved from http://acbio.org.za/wp-content/uploads/2015/12/Seed-Sector-Sub-Sahara-report.pdf

ACB (African Centre for Biosafety). (2014). *The political economy of Africa's burgeoning chemical fertiliser rush*. Retrieved from http://acbio.org.za/wp-content/uploads/2014/12/Fertilizer-report-201409151.pdf

Africa, V. (2014). *Massmart takes on Shoprite as it expands into West African retail market*. Retrieved from http://venturesafrica.com/massmart-takes-on-shoprite-as-it-expands-into-west-african-retail-market/

Africa Business Journal. (2014). Tiger brands shows its stripes with aggressive expansion. *Africa Business Journal* 5(6). Retrieved from http://www.tabj.co.za/food_drink/the_tiger_brands.html

AgriSA Africa Policy Committee (2014). *Annual report* 2013–2014. Retrieved from http://www.agrisa.co.za/pdf/AGRI%20SA%20Annual%20Report%202013-14%20Part%2002%20-%20Policy%20Committee.pdf

Aliber, M., Maluleke, T., Manenzhe, T., Paradza, G., & Cousins, B. (2013). *Land reform and livelihoods: Trajectories of change in Limpopo province, South Africa*. Human Sciences Research Council: Cape Town.

Anseeuw, W., Boche, M., Breu, T., Giger, M., Lay, J., Messerli, P., & Nolte, K. (2012). Transnational land deals for agriculture in the Global South: Analytical report based on the Land Matrix database. Bern/Montpellier/Hamburg. CDE/CIRAD/GIGA.

Bernstein, H. (1996). The agrarian question in South Africa: Extreme and exceptional? *Journal of Peasant Studies*, 23(2-3), 1–52.

Bernstein, H. (2013). Commercial agriculture in South Africa since 1994: 'Natural, simply capitalism'. *Journal of Agrarian Change*, 13(1), 23–46.

Bhorat, H., Buthelezi, M., Chipkin, I., Duma, S., Mondi, L., Peters, C., & Friedenstein, H. (2017). *Betrayal of the promise. How South Africa is being stolen*. Centre for Complex Systems in Transition (University of Stellenbosch), Public Affairs Institute (University of, Witwatersrand), Development Policy Research Institute, University of Cape Town), South African Research Chair in African Diplomacy and Foreign Policy (University of Johannesburg).

Bhorat, H., Hirsch, A., Kanbur, R., & Ncube, M. (2014). Economic policy in South Africa – past, present and future. In H. Bhorat, A. Hirsch, R. Kanbur, & M. Ncube (Eds.), *The Oxford companion to the economics of South Africa* (pp. 1–25). Oxford: Oxford University Press.

Boche, M., & Anseeuw, W. (2013). *Unraveling 'land grabbing' – different models of large-scale land acquisition in Southern Africa* (Land Deal Politics Initiative Working Paper 46). Retrieved from http://www.plaas.org.za/sites/default/files/publications-pdf/LDPI46Boche%26Anseeuw.pdf

Ciuri, S. (2014, July 27). Nakumatt takes over first ShopRite store. *Business Daily*. Retrieved from http://www.businessdailyafrica.com/corporate/Nakumatt-takes-over-first-Shoprite-store/539550-2399768-8pdjjq/index.html

Cock, J. (2009). Declining food safety in South Africa: Monopolies on the bread market. In *The global crisis and Africa: Struggles for alternatives* (pp. 1–4). Johannesburg: Rosa Luxemburg Stiftung.

Coleman, A. (2012, August 1). The Pannar-Pioneer merger explained. *Farmer's Weekly*. Retrieved from http://www.farmersweekly.co.za/article.aspx?id = 26340&h = The-Pannar-Pioneer-merger-explained

Cousins, B. (2013). Land reform and agriculture uncoupled: The political economy of rural reform in post-apartheid South Africa. In P. Hebinck, & B. Cousins (Eds.), *In the shadow of policy. Everyday practices in South Africa's land and agrarian reform* (pp. 47–62). Johannesburg: Wits University Press.

Cousins, Ben and Ruth Hall. (2013). Rights without illusions: The potential and limits of rights-based approaches to securing land tenure in rural South Africa. In M. Langford, J. D. Ben Cousins, & T. Madingozi (Eds.), *Symbols or substance? The role and impact of socio-economic rights strategies in South Africa* (pp. 157–186). Cambridge: Cambridge University Press.

Cousins, B., & Walker, C. (Eds.) (2015). *Land divided, land restored: Prospects for land reform in 21st century South Africa*. Cape Town: Jacana.

'Diyan, Tonia. (2016 January 16). Shoprite Nigeria procures 76% of products sold locally. *The Nation*. Retrieved 29 November 2017 from http://thenationonlineng.net/shoprite-nigeria-procures-76-of-products-sold-locally/.

Douglas, K. (2014, January 7). 'What does the failure of Woolworths say about Nigeria?' How we made it in Africa website. Retrieved from https://www.howwemadeitinafrica.com/what-does-the-failure-of-woolworths-say-about-nigeria/

Dubb, A. (2016). Interrogating the logic of accumulation in the sugar sector in Southern Africa. Journal of Southern African Studies, 43(3), 471–499.

Emongor, R. (2008). Namibia: Trends in growth of modern retail and wholesale chains and related agribusiness. Policy brief 8, regoverning markets. Pretoria: University of Pretoria.

Fairbairn, M. (2015): Foreignization, financialization and land grab regulation. Journal of Agrarian Change, 15 (4), 581–591.

Fine, B. (2008, April 4–6). 'The Minerals-Energy Complex is dead: Long live the MEC?' Paper presented at the Amandla colloquium on 'Continuity and discontinuity of capitalism in post-apartheid South Africa', Cape Town.

Goko, C. (2016, April 28). 'Pick n Pay to enter Nigerian market.' Business Day. Retrieved from http://www.bdlive.co.za/business/retail/2016/04/28/pick-n-pay-to-enter-nigerian-market

Gordhan, P. (2013, February 27). Budget speech: Minister of finance, Pravin Gordhan. Government of the Republic of South Africa. Retrieved from http://www.treasury.gov.za/documents/national%20budget/2013/speech/speech.pdf

Govender, G. (2013). Financing for development? The development bank of South Africa and its footprint in Africa. Internal discussion document. Johannesburg: Action Aid.

Greenberg, S. (2017). Corporate power in the agro-food system and the consumer food environment in South Africa. Journal of Peasant Studies, 44(2), 467–496.

Hall, R. (2011). Land grabbing in Southern Africa: The many faces of the investor rush. Review of African Political Economy, 38(128), 193–214.

Hall, R. (2012). The next great trek? South African commercial farmers move north. Journal of Peasant Studies, 39(3&4), 823–843.

Hall, R., Anseeuw, W., & Paradza, G. (2015). South African commercial farmers in the Congo. In R. Hall, I. Scoones, & D. Tsikata (Eds.), Africa's land rush: Rural livelihoods and agrarian change (pp. 162–180). Oxford: James Currey.

Hall, R., & Kepe, T. (2017). Elite capture and state neglect: New evidence on South Africa's land reform. Review of African Political Economy, 44(151), 122–130.

Hammar, A. (2010). Ambivalent mobilities: Zimbabwean farmers in Mozambique. Journal of Southern African Studies, 36(2), 395–416.

Harding, C. (2011). Nakumatt pushes further into Africa. Should ShopRite be worried? How we made it in Africa. Retrieved from http://www.howwemadeitinafrica.com/nakumatt-pushes-further-into-africa-should-shoprite-be-worried/14026/

Harvey, D. (2003). The new imperialism. Oxford: Oxford University Press.

Independent On Line (IOL). (2013, February 4). Rainbow to buy stake in Zam Chick. Retrieved from http://www.iol.co.za/business/companies/rainbow-to-buy-stake-in-zam-chick-1463525

Jayne, T. S., Chamberlin, J., & Headey, D. D. (2014). Land pressures, the evolution of farming systems, and development strategies in Africa: A synthesis. Food Policy, 48, 1–17.

Marais, H. (2011). South Africa pushed to the limit. The political economy of change. Cape Town: UCT Press.

Massmart. (2012). Massmart in Africa: May 2012 review. Group update. Retrieved from http://www.massmart.co.za/wp-content/uploads/2013/12/Massmarts_operations_africa_may2012_review_v5.pdf

Matrix, L. (2015). Retrieved from http://www.landmatrix.org/en/get-the-detail/by-investor-country/south-africa/?order_by=&more = 90

McNellis, P. (2009). Foreign investment in developing country agriculture: The emerging role of private sector finance (Commodity and Trade Policy Working Paper 28). Rome: Food and Agricultural Organization of the United Nations.

Merian Research and CRBM (Campaign for the Reform of the World Bank). (2010). The vultures of land grabbing: the involvement of European financial companies in large-scale land acquisition abroad. No publication details. Retrieved from http://www.rinoceros.org/IMG/pdf/VULTURES-completo-2.pdf

Moore, J. (2015). Capitalism in the web of life: Ecology and the accumulation of capital. London: Verso Books.

PAFO (Pan African Farmers' Organization). (2014, December 12). 'Africa must take ownership of its agriculture' – says newly elected PAFO President. Retrieved from http://pafo-africa.org/africa-must-take-ownership-of-its-agriculture-says-newly-elected-pafo-president/

Peacock, B. (2016, June 23). Danone and Clover face off on the shelves. Business Day. Retrieved from http://www.bdlive.co.za/world/europe/2016/06/23/danone-and-clover-face-off-on-the-shelves

Pioneer Foods. (2014). Summary consolidated financial statements for the year ended 30 September 2014. Retrieved from http://www.pioneerfoods.co.za/2014/11/summary-consolidated-financial-statements-for-the-year-ended-30-september-2014/

Pitso, R. (2016, June 23). Poultry producers look beyond SA for growth. Business Day. Retrieved from http://www.bdlive.co.za/business/agriculture/2016/06/23/poultry-producers-look-beyond-sa-for-growth

Radebe, Sibonelo. (2016, August 12). South Africa is Africa's largest economy again – but what does it mean? The Conversation. Retrieved from http://theconversation.com/south-africa-is-africas-largest-economy-again-but-what-does-it-mean-63860

RCL Foods. (2016a). Our African partnerships. Retrieved from http://www.rclfoods.com/ourafricanpartnerships

RCL Foods. (2016b). Zam Hatch Making RCL Foods History. Retrieved from http://www.rclfoods.com/zamhatch

Scoones, I., Smalley, R., Hall, R., & Tsikata, D. (2014). Narratives of scarcity: Understanding the 'global resource grab' (FAC Working Paper 62). Brighton: Future Agricultures Consortium. Retrieved from http://www.future-agricultures.org/publications/research-and-analysis/1830-narratives-of-scarcity-understanding-the-global-resource-grab/file

Shevel, A. (2014, March 2). Shoprite cashes in on underestimated Africa. Business Day. Retrieved from http://www.bdlive.co.za/africa/africanbusiness/2014/03/02/shoprite-cashes-in-on-underestimated-africa

Shoprite Holdings Limited. (2016). Geographical spread. Retrieved from http://www.shopriteholdings.co.za/OurGroup/Pages/Geographical-Spread.aspx

Smalley, R., Sulle, E., & Malale, L. (2014). The role of the state and foreign capital in agricultural commercialisation: The case of sugarcane outgrowers in Kilombero District, Tanzania (Working Paper 106). Brighton: Future Agricultures Consortium. Retrieved from http://www.plaas.org.za/sites/default/files/publications-pdf/FAC_Working_Paper_106.pdf

UFF. (2015). UFF agri asset management: Company profile. Retrieved from http://www.uff.co.za/CorporateProfile-3450.html

Unitrans. (2017). Unitrans company website. Retrieved from https://www.unitrans.co.za/

Vink, N., & Rooyen, J. v. (2009). The economic performance of agriculture in South Africa since 1994: Implications for food security. Halfway House: Development Bank of Southern Africa.

Walker, C., Bohlin, A., Hall, R., & Kepe, T. (Eds.). (2010). Land, memory, reconstruction and justice: Perspectives on land claims in South Africa. Athens, OH: Ohio University Press.

Weatherspoon, D., Neven, D., Katjiuongua, H. B., Fotsin, R., & Reardon, T. (2004). Battle of the supermarket supply chains in sub-Saharan Africa: challenges and opportunities for agrifood suppliers. Geneva: United Nations Conference on Trade and Development. Retrieved from http://unctad.org/en/Docs/ditccommisc20035_en.pdf

Weatherspoon, D., & Reardon, T. (2003). The rise of supermarkets in Africa: Implications for agrifood systems and the rural poor. Development Policy Review, 21, 333–355.

Who Owns Whom. (2017a, February 28). Manufacture of Pesticides and Other Agrochemical Products. Retrieved from https://www.woweb.co.za/?m = Industries&p = reportinfo&id = 4486&country = 222&tab = 8

Who Owns Whom. (2017b, February 24). Wholesale and Retail of Food in Tanzania. Retrieved from https://www.woweb.co.za/?m = Industries&p = reportinfo&id = 4487&country = 235&tab = 8)

Wolpe, H. (1972). Capitalism and cheap labour power: From segregation to apartheid. Economy and Society, 1 (4), 425–456.

The ambiguous stance of Brazil as a regional power: piloting a course between commodity-based surpluses and national development

Sérgio Sauer, Moisés V. Balestro and Sergio Schneider

ABSTRACT

This article argues that economic growth relying on commodity-based exports – combined with domestic market expansion for consumption, an overvalued exchange rate and high interest rates – constrained national development as much as it did Brazil's status as a regional power, particularly in the 2000s. The Brazilian 'neo-developmentalist model' – pursuing export surplus in the balance of trade and foreign investment – was based excessively on government incentives for the export of natural resources. With regard to agrarian issues, Brazil again played an uncertain role as a regional power during the governments of both Lula (2003–2010) and Rousseff (2011–2016), despite important differences between the two administrations. On the one hand, the country encouraged the transfer of family farming policies to other Latin American countries. On the other hand, the government's 'national champions policies' were also paramount in forging the expansion of agribusinesses and other multinationals across the continent. The very nature of this ambiguity rests on the contradictions between the narrative of a national development project and the reality of deindustrialization and commodity-based economic surplus. By drawing on aggregate data and secondary sources, this article explores the limits and contradictions of the Brazilian development path in becoming a more influential regional power in the 2000s.

Introduction

The global and regional influence of the BRICS (Brazil, Russia, India, China, and South Africa) countries has increased. As BRICS moved beyond the Goldman Sachs acronym and began to grow as a co-operative, as well as competitive, group of countries, their regional influence started to draw attention. These countries are rising regional powers, and the inclusion of South Africa, commonly referred to as the gateway to Africa, in 2011, follows this trend. BRICS has truly unveiled its geopolitical reach in a multipolar world order (Carpintero, Murray, & Bellver, 2016; Costa Leite et al., 2014), combining global co-operation and regional competition. While this is a crucial dimension, the article's main arguments are related not to the general influence of BRICS in Latin America, but only to Brazil and its recent policies and narratives as a regional power.

Prior to BRICS, Brazil's prominent role in the southern cone of South America can be traced back to the creation of Mercosur in 1991, or even before that. However, most recently, the country's position was reinforced by the reactivation of agreements within Mercosur as part of a shift in Brazilian foreign policy within the Workers' Party governments. Since 2003, there had been greater emphasis in its foreign policy regarding the southern hemisphere on so-called South–South co-operation. Also, the founding of the Union of South American Nations in 2006 not only strengthened South–South co-operation, but also underlined the Brazilian state's influence (Kellogg, 2007), which has been under threat since 2016, when Michel Temer took control of the government during the ongoing economic crisis.

The wave of progressive, left-leaning governments in South America contributed to a more prominent role for Brazil as a regional power in the first decade of the twenty-first century (Vergara-Camus & Kay, 2017). Strengthening unity and co-operation in Latin America were essential in the narratives of these governments. This has been particularly true for regional co-operation and agreements intended to enhance areas of common interest such as infrastructure, energy, farming and, to a lesser extent, labour regulations (Curado, 2015). At the same time, Latin American economic growth was fuelled by commodity and mining exports catering for Chinese demand (McKay, Alonso-Fradejas, Brent, Sauer, & Xu, 2016; Wesz, 2016; Wilkinson, Wesz, & Lopane, 2016). Trade between Latin America and China has experienced a 22-fold increase since 2000, and Chinese public finance has become the largest creditor, lending US$125 billion to the region between 2005 and 2015 (McKay et al., 2016). Between 2001 and 2010, mining and fossil fuel exports from Latin America grew at 16% annually, according to the Organisation for Economic Co-operation and Development (OECD, 2016a). Nevertheless, this growth may run out of steam due to a lower Chinese demand for commodities, in spite of a 2015 pledge by President Xi Jinping to invest US$250 billion in the region (McKay et al., 2016).

The ambiguous nature of progressive governments in Latin America, particularly the PT governments in Brazil, is reflected in the narratives of South–South co-operation and the development project, in contrast to the deindustrialization and reprimarization of the Brazilian economy (Cano, 2012; Jenkins, 2015; Jenkins & Barbosa, 2012; Rodrik, 2016; Veltmeyer, 2017). Such ambiguity between a desired political and economic project and the reality of deindustrialization and a higher dependency on the export of commodities undermines the continuity of progressive political elites (Sauer & Mészarós, 2017; Vergara-Camus & Kay, 2017). Recent political changes throughout the continent, such as the parliamentary coup in Brazil in 2016 to implement neoliberal reforms and the electoral victory of a neoliberal government in Argentina in 2015 corroborate this claim.

The diffusion of policies supporting family farming and the strengthening of large agribusiness enterprises expresses the ambiguous nature of Brazilian national neo-developmentalism as much as it does its influence as a regional power. On the one hand, the country has encouraged the dissemination of policies (Sabourin, Samper, & Sotomayor, 2015) on rural territorial development and food acquisition programmes among other Latin American countries. Positive experiences in the areas of family farming and social policy are mostly related to action undertaken by the Brazilian Co-operation Agency (ABC), the now-dissolved Ministry of Agrarian Development, the Ministry of Social Development and the Brazilian Agricultural Research Corporation (EMBRAPA). On the other hand, the 'national champions policy' (BNDES, 2017) implemented during the Lula (2003–2010) and, less so, the Rousseff (2011–2016) administrations reinforced the presence of the Brazilian agribusiness multinationals. As the supply chain of these multinationals is built around large food processors, these national champions encouraged the spreading of agribusiness enterprises to other Latin American countries. The world's largest meat processor, the Brazilian JBS Group, is a prime example of the strength of agribusiness (Moraes, 2017). Brazilian agribusiness has spread to both middle-income countries such as Argentina and Uruguay, and low-income countries such as Paraguay and Bolivia.

Furthermore, the actions taken by the Brazilian state are shaped by the interests of large business groups much more than by policy networks for family farming. The loans extended and investments made by the National Bank for Economic and Social Development (BNDES, 2017) were directed to large Brazilian companies in agribusiness, mining, construction, and infrastructure. In addition, there were significant stock purchases from pension funds in these companies.

To shed light upon the ambiguous and shaky rise of Brazil as a regional power, the article has four main sections. The first section provides an overview of the recent trends in Latin America, bringing together aggregate data from the last decade, trying to demonstrate that the economic growth has been based almost exclusively on exporting raw materials. Brazil's public support for such development illustrates its neo-extractivist character. The second section discusses the political role of Brazil as a regional power in Latin America. Such a role is based mainly on the South–South, or even regional, co-operation narratives, and much less so on direct investment in the region. The third section discusses Brazilian influences in policy-making, reproducing its own internal contradictions, especially its public backing for the internationalization of large agribusiness firms, but also its encouragement for the reproduction of family farming policies and programmes. In the fourth section, Brazil is analysed as a regional power, including a consideration of its recent co-operation practices and projects.

The main argument asserting the ambiguous and shaky position of Brazil as a regional power has to do with the contradictions stemming from three trends. One is the resumption of a neo-extractivist economic model based on the export of commodities. The second is deindustrialization (or reprimarization), which entails a diminishing role for the country in global value chains and manufacturing. The third trend is the continuity of an overvalued currency and extremely high interest rates. The combination of these trends hinders the emergence of Brazil as a regional power.

Economic growth, reprimarization, and poverty reduction in Latin America

The commodities boom during the first decade of the twenty-first century helped several Latin American countries to achieve substantial trade surpluses without adding value to their exports. Rising prices of commodities were also the base for recent Brazilian economic growth and its international trade surplus. Based on strong governmental support (given its neo-developmentalist perspective), such positive trade balance was achieved by exporting primary commodities, generating revenue to sustain the government's social programmes and to reduce poverty (in accordance with its neo-extractivist logic).

The average GDP growth in Latin America was 2.86% between 2005 and 2013, which was slightly above world GDP growth of 2.52%. Foreign direct investment (FDI) also grew from US$57.8 billion in 2005 to US$157.5 billion in 2013. Together with economic growth, Latin America succeeded in reducing poverty levels from 48.4% in 1990 to 28.1% in 2013 (ECLAC, 2014). Reduction in poverty levels and increased food security were associated with progressive governments, such as those of Kirchner in Argentina, Morales in Bolivia, Correa in Ecuador, and Lula in Brazil. The political economy of these governments is loosely and vaguely associated with 'neo-developmentalism' (Boito & Berringer, 2014) and/or 'neo-extractivism' (Baletti, 2014; Veltmeyer & Petras, 2014).

In the case of the neo-extractivism, Burchardt and Dietz (2014) mention three main features of the state: (a) it regulates the economies of extraction and mediates between diverging interests from different social groups; (b) it acts as an agent of development and addresses public-interest issues by supporting development projects; and (c) it creates political legitimacy through democratic elections and a development narrative.

In particular, these features characterize the case of Brazil, the economic model of which allowed for the creation and/or expansion of public policies and (direct cash transfer) social programmes over the last decade (Baletti, 2014; Sauer & Mészarós, 2017).

Nevertheless, like other regions around the world, Latin America's economic growth began to stagnate as the effects of the 2008 global crisis appeared. Progressive governments and social movements throughout the region shared a broad concern about the maintenance of social gains – especially social inclusion, poverty reduction, and food security – when the commodity boom started to lose momentum. There is a growing risk that social policies will be undermined by sluggish growth within an economy based on a neo-extractive model. This risk triggers the debate on deindustrialization and reprimarization of the Latin American economy (De la Torre, Didier, Ize, Lederman, & Schmukler, 2015; Jenkins, 2015).

Except for Mexico, where the assembly of manufactured goods for export is the engine of the economy, manufacture in major economies in Latin America has lost momentum over the last decade. Of these economies, Brazil suffered the largest loss of GDP share in manufacturing. Compared with China, where manufacturing accounts for more than 30% of GDP, Brazilian manufacturing accounts for little more than 10% of GDP (see Figure 1). As a latecomer in the catching-up process, this premature loss of manufacturing strength spells bad news for the country's economic and social development. There was also a substantial decrease in manufactured goods as a share of overall Brazilian exports between 1994 and 2012, from 63.65% to 39.08% (MDIC, 2015).

Recently, there have been efforts to move towards regional integration by increasing trade between Latin American countries. However, such trade has been mainly in consumer rather than capital goods (Cervo & Lessa, 2014). In 2016, trade in certain capital goods and intermediary goods dropped by 7% and 10% respectively (CEPAL, 2016). Latin American exports of capital goods to Brazil plunged by 45% between 2015 and 2016.

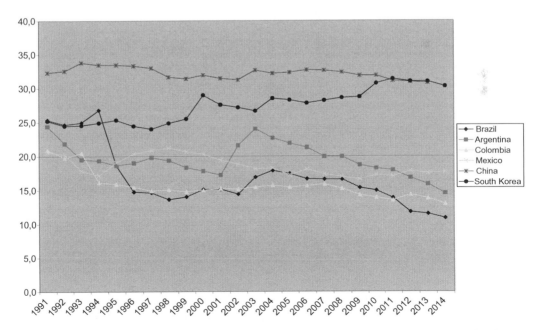

Figure 1. Manufacturing value-added share (% of GDP), 1991–2014. Source: World Bank – databank.worldbank.org/data/home.aspx.

This drastic fall has two major implications. One is its adverse effect on the current account surplus, because there is a positive correlation between export surplus in capital goods and the current account (CEPAL, 2016). A country with a lower current account surplus is more dependent on foreign savings to finance the economy, and the upshot of this is an overvalued currency, which negatively affects the export of manufactured goods (Bresser-Pereira, 2015). The second implication is a greater reliance upon natural resources extraction and commodity production as a source of economic growth.

In addition, Latin American countries have increased their volume of extra-regional trade with China. With the loss of manufacturing capacity in major Latin American economies, unfavourable terms of exchange with China arise. Of the goods that China imported from Brazil in 2011, 83.6% were 'basic products' and only 4.7% were 'manufactured products'. This reveals an association between China's rise and the reprimarization of Latin American countries' exports (Curado, 2015).[1]

Considering the aggregate exports from Latin America and the Caribbean (LAC) (see Figure 2), only 2% of manufacturing exports go to China. On the other hand, exports of minerals and crops, in the extractive and agricultural sectors, account for 30% of the total. While China is certainly one of the main destinations for raw material exports from LAC, it is one of the least important destinations for manufacturing exports.[2]

This is harmful to domestic-based industrial elites who manufacture intermediary goods, equipment, and durable consumer goods. In Brazil, these elites have less bargaining power in decisions regarding macroeconomic policy. They tend to organize in industry business associations such as the Brazilian Business Association of Machinery and Equipment Manufacturers and the Brazilian Business Association of the Electronic Industry. In general, they have much less influence and power than agribusiness groups like huge retail chains (and supermarkets), agribusiness firms, and farms (Schneider, 2009, 2012). Large family business groups are concentrated in low-added value and technologically mature types of industry, as in the case of civil construction and food and beverages. They profit more from domestic markets and less-developed markets in low-income Latin American countries in such a way that they are less affected by commodity-based exports and the import of value-added consumer goods (Schneider, 2013).

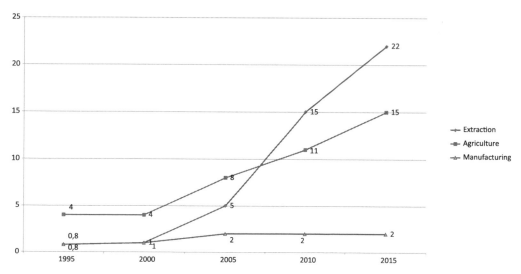

Figure 2. China's share of LAC exports by sector, 1993–2013. Source: Ray and Gallagher (2017).

In the Brazilian case, primary products and semi-finished goods (minerals, but also agricultural commodities) receive almost total tax exemption under the Kandir Act (Complementary Law 87), issued in 1996.[3] On the other hand, exporters of manufactured goods have more difficulty in obtaining tax exemptions. In addition to the demand for commodities on the global market, such a tax system favours the export of primary products and raw materials at the expense of manufactured goods (BNDES, 2014). It shows a combination of internal advantages and international business opportunities reviving the agro-export model and neo-extractivism that reigned under the 'commodity boom' of the last decade.

Consequently, the Brazilian balance of trade is heavily dependent upon the export of primary commodities. Brazil has been showing a positive balance of trade since 2001, with agricultural exports representing a constant share. These exports accounted for an average of 42% of total exports from 1999 to 2010, and 39.5% in 2012 (Conceição & Conceição, 2014). Soybean exports were responsible for 23.2%, and soy feed for 8.8% of total agricultural exports in 2012. While there are also exports of sugar (13.3%), chicken meat (9.6%), and beef (7.6%), the soybean complex remains the chief exporter. This restricted export list had China as its principal destination, with 18.8% of the total (as against 5.5% in 2002), and the United States with 7.3%in 2012 (16.7% in 2002) (Conceição & Conceição, 2014).

Thus, large national and multinational agribusiness firms profit from a stronger Chinese demand for commodities, as shown in Figure 2. In turn, financial capital benefited most from government bonds with skyrocketing interest rates and the surge of speculative investments such as those associated with real estate and the acquisition of land (Freitas, 2011).

The positive trade balance, which has resulted from the rising prices of commodities over more than a decade, has served to lessen the effects of these unequal terms of trade with China. As stated by Salama (2013), the new primarization of exports has taken advantage of the terms of the trade balance in such a way that foreign constraints are softened and vulnerability to this commodity-export dependency is reduced.

The growth of a trade surplus meant that more foreign currency was being pumped into the Brazilian economy and there was more revenue to sustain the government's social and pro-business policies. In the case of Brazil, the amount of revenue directed to family farming, conditional cash transfer programmes (*Bolsa Família*), and the unemployment insurance programme increased three-fold during the Lula and Dilma governments (Boito & Berringer, 2014).

As Jenkins (2015) puts it, rather than being attributed solely to Chinese import competition requiring protectionist measures, the problem of deindustrialization must be seen as highlighting the nature of Brazil's role in the global economy (agro-export model). Furthermore, it needs a competitive exchange rate and industrial policies that could effectively contribute to changing the productive structure, increasing the share of middle- and high-technology industries.

Returning to the recent role of China in the region, Figure 3 shows that it is not only a matter of importing value-added goods in exchange for the export of commodities (McKay et al., 2016). When the distribution of FDI flows from China is considered, mature industries and natural resources account for the largest share. Latin America accounted for 13% of Chinese FDI in 2011 (ECLAC, 2013). That represents twice the amount of Chinese FDI in Europe (5.8%) and three times that in Africa (3.8%). Items such as food, beverages, tobacco, and extraction account for 55% of Chinese FDI flowing through the region (Ray & Gallagher, 2017).[4] Not only imports, but also Chinese investments reinforce the primarization trend.

Latin America also stands out among the world regions that have achieved a reduction in hunger and poverty, meeting the Millennium Development Goals. According to Cervo and Lessa (2014), the

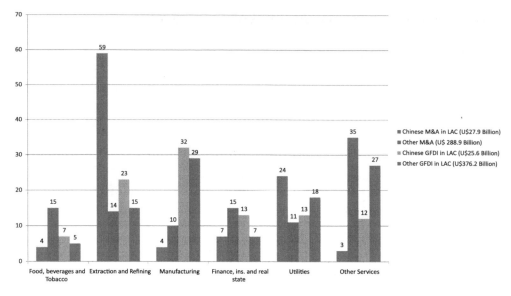

Figure 3. Distribution from Chinese FDI flows to LAC. Source: Ray and Gallagher (2017).

regional governments' agreements aim to overcome asymmetries and harmonize productive integration and investment in infrastructure in the fight against poverty and for social inclusion. The virtuous combination of economic growth, political and institutional stability, and incentives for agriculture and rural development were highlighted in a recent report by the Food and Agriculture Organisation (FAO), International Fund for Agricultural Development (IFAD) and World Food Programme (WFP) (2013) on the state of food security.[5]

According to the report (FAO, IFAD & WFP, 2013), LAC as a sub-region, and South America in particular, achieved the adequate supply of food decades ago when production exceeded consumption. However, the report added that such growth had not been sufficiently inclusive to ensure access to food for all. The report claims that economic growth alone is not enough to ensure sustainable food security and nutrition due to a high level of inequality and substantial numbers of low-paid workers.

Despite consistent improvements in poverty reduction, further positive results in the region's social development were upset by the end of the commodity boom, which was accompanied by a significant fall in export prices. For latecomers, such as the more developed LAC countries, the premature reduction in the manufacturing share of GDP and greater reliance on primary products tend to hamper sustained levels of social development with the creation of more skilled and semi-skilled jobs.

In summary, Brazil experienced significant growth rates in the first decade of the twenty-first century (for example, with GDP growth rates of 5.7% in 2004, 6.0% in 2007, and 7.5% in 2010), mainly exporting primary commodities and providing incentives for domestic consumption. On the one hand, even with an international trade surplus, Brazilian trade – as it has been historically, and unchanged by BRICS or in exchanges with China – ended up reinforcing the reprimarization trend. On the other hand, under progressive governments and their neo-developmentalist policies, portions of the revenue were shared through social programmes – in addition to increases in labour income or wages – contributing to reduced rates of poverty.

Brazil's aspirations to become a regional power

Considering the country's economy and size, Brazil was historically (and undoubtedly) a regional power. However, its leadership took a new direction after 2003, mainly because of a combination of three different factors: (a) a new level of economic growth in Brazil; (b) the election of progressive governments in several South American countries, which sought regional co-operation based on South–South co-operation narratives, including ways to avoid US neoliberal influences; and (c) the concept of foreign co-operation as solidarity, and the emergence of BRICS. In the light of these factors, the Lula administration launched its 'national champions policy' to support the internationalization of Brazilian firms, as part of Brazil being or becoming a regional power.

Despite its strength on the continent, Brazil has been historically and politically subordinate to the geopolitical interests of the United States (Marini, 1973), especially after the Second World War and the import-substitution period that lasted until 1980. During the 1980s and 1990s, the return of democracy to Latin America coincided with the rise of the neoliberal Washington Consensus. The depth and intensity of the pro-market reforms were different in Latin American countries. There were more widespread effects in the case of Argentina, Colombia, and Mexico, in comparison to Brazil. In reaction to the neoliberal policies, the election of progressive governments contributed to a leading role for Brazil based on the narrative of regional co-operation.

Besides that, following the macroeconomic stabilization in Brazil that came with Cardoso's Real Plan in 1994, there were no disruptive crises, neither political nor economic, between 1994 and 2014. This stability helped to create conditions that allowed for President Lula's strategy to strengthen Brazil as a regional power that would play a prominent role in international organizations and further the agenda of South–South co-operation (IPEA, 2010).

President Lula's two terms (2003–2010) focused attention on South America through policies directed at specific countries, such as Bolivia, Paraguay, and Venezuela. Ideological and political concerns generated close attention to the South American continent. Although they did not always yield the expected results, the policies pursued and resources invested in this region served to enhance Brazil's influence. Such policies and aid included writing off Bolivia's debt (US$52 million in 2004) and investments by government agencies such as the Bank of Brazil (*Banco do Brasil*) and the BNDES (2017). These public banks financed projects on the continent and the establishment of the Initiative for the Integration of Regional Infrastructure of South America (IIRSA) after 2000 (Iglesias, 2008; Safransky & Wolford, 2011).[6]

Beyond the rhetoric characterizing Brazil as a large country with enormous potential to become a global player, there are two dimensions to be considered. One is the growing economic importance of Brazilian investments abroad, and the other has to do with the country's growing role in international co-operation, especially among countries in the southern hemisphere.[7]

Nevertheless, 32.2% of total Brazilian FDI was in tax havens, and 7.44% was in Latin America in 2012. In fact, Brazilian FDI in Latin America decreased from 13.76% in 2001 to 7.44% in 2012. During the same period, tax havens saw a decrease in their share of FDI, from 68.2% to 32.2%. However, the share in the European Union increased from 8.42% in 2001 to 50.26% in 2012 (CNI, 2013). Of the 50.26% of Brazilian FDI in the European Union, 42% was concentrated in five countries with tax havens. The largest part remained in Austria (23%), the Netherlands (11.4%), and Luxemburg (6%) (CNI, 2013). FDI stocks in tax havens do not contribute to the country's political and economic power.

There are two major reasons for the overwhelming presence of this speculative type of investment. One is the fact that the rentiers are extremely powerful in Brazil, being the strongest group within the economic elite. These rentiers consist of institutional investors, private banks, and various large economic groups (such as Globo, the giant media company). Their main investment is in treasury bonds, a form of Brazilian public debt security, but also housing funds, the acquisition of land and derivatives (Freitas, 2011).[8] The second reason is the gradual transformation of sections of the Brazilian industrial elite into financial speculators, and importers of capital and intermediary goods.

Within capitalist dynamics, the accumulation of capital abroad in the form of productive investments in manufacturing and services is a key element in strengthening the global presence of a country. At the same time, this process is incontrovertibly political in the sense that state policies constrain and shape the behaviour of indigenous firms in host countries as well as in their home countries. According to Spar (2001), as long as nation states mediate the struggle over resources to leverage firms' capabilities, they remain influential actors on the stage of global business (Figure 4).

In addition, as shown in Figure 5, the uneven and significant rise of Brazilian FDI in the first decade of the twenty-first was followed by a sharp decline after 2010. The BNDES certainly played a role in the expansion of Brazilian FDI by strongly encouraging the internationalization of large companies in an attempt to catalyse the formation of Brazilian multinationals in industries where the country has a competitive edge, such as agribusiness,[9] mining, and the construction sector. However, this strategy, called the 'national champions' policy (BNDES, 2017), was abandoned by BNDES in 2012 (Valor, 2012). The main criticism against the idea of a policy to build national champions came from Lazzarini and Musacchio (2014), who claimed that BNDES loans and acquisition of shares merely transferred subsidies to big business without any marked benefit to the firms' performance levels. Also, economists and the press considered taxpayers' money should be used more efficiently.[10]

The main types of industry making up the structure of Brazilian FDI are commodity-based and financial services. Oil and gas industries, with massive international investments by PETROBRAS,

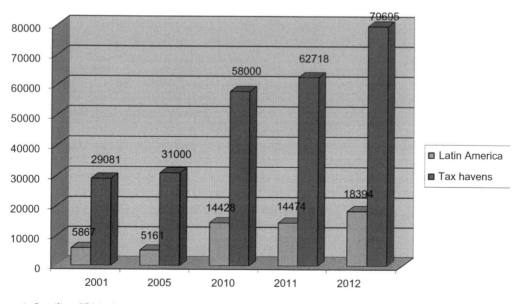

Figure 4. Brazilian FDI in Latin America and in tax havens, 2001–2012 (US$ million). Source: CNI (2013).

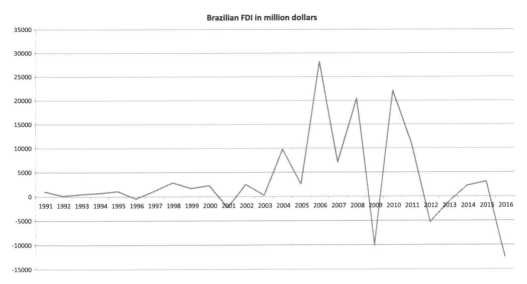

Figure 5. Rise and fall of Brazilian FDI, 1991–2016. Source: UNCTADSTAT (2017).

accounted for US$12.4 billion, but agriculture, livestock, and husbandry are responsible for nearly US$62 billion of Brazilian FDI (see Table 1). The proportion of financial services in total FDI jumped from US$38 billion to U$98 billion between 2007 and 2012. The third-largest industry, metallic minerals, increased from US$36.4 billion in 2007 to 42.8 billion in 2012 (CNI, 2013).

Undoubtedly, Brazilian agribusiness, underpinned by international and national capital, played a pivotal role in the expansion of the country's investments in Latin America over the past decade (Moraes, 2017). Regarding productive FDI, as shown in Table 1, agriculture and livestock rank second in Brazilian FDI, increasing from US$37.9 billion in 2007 to US$61.9 billion in 2012.

As a driver of regional integration, Brazil has had an important role in IIRSA, which has a portfolio of 510 infrastructure projects totalling US$74.5 billion. The funding of IIRSA is supplied mainly by national governments (62%) and development banks (20%). Thus far, the chief purpose of the infrastructure projects has been the creation or improvement of export corridors (Safransky & Wolford, 2011). A major example is the Peru-Brazil-Bolivia Hub, consisting of 26 projects organized into three groups, with an estimated investment of US$29.1 billion (Iglesias, 2008). Of this amount, 62.6% will go into the Brazilian project known as the Madeira River Hydroelectric Power Complex (Santo Antônio and Jirau Hydroelectric Dams), which is currently being built.[11]

Table 1. Brazilian FDI by selected types of industry (US$ million).

Type of industry	2007	2008	2009	2010	2011	2012
Agriculture and livestock husbandry	37,917	35,324	41,855	52,855	61,649	61,936
Financial services	37,785	37,098	46,325	64,128	74,284	97,928
Metallic minerals	36,382	33,897	39,947	44,964	49,164	42,806
Food and beverages	9165	7714	10,820	8572	6114	16,928
Steel	2858	2964	4119	8490	14,730	15,173
Oil and natural gas production	1431	1322	1432	6323	8554	12,348
Chemical products	56	37	501	511	611	731
Automotive vehicles	341	560	569	646	606	596
Machinery and equipment	141	154	253	322	348	229

Source: CNI (2013).

Since 2011, BNDES' investments have surpassed those of the World Bank's loans in the region in several countries such Argentina, Chile, Venezuela, Ecuador, Cuba, Colombia, and Uruguay). Brazil claims an 18% share of engineering services in the Latin American and the Caribbean market. The major infrastructure projects being funded by the BNDES are in Cuba, Colombia, Argentina, Ecuador, and Peru (see Table 2).

Supporting the internationalization of Brazilian business, BNDES' first international operation took place in 2005. The bank's external activities have mainly focused on supporting Brazilian firms taking part in mergers and acquisitions overseas (Doctor, 2015). The creation and internationalization of Brazilian multinationals were clearly supported by BNDES loans and equity financing, as part of the strategy for increasing ' … economic efficiency and the development and diffusion of technologies with greater potential to induce the level of activity and competition in the foreign market' (BNDES, 2017, p. 135). BNDES' financial support targeted ' … sectors with international projection capabilities: Bioethanol; Petroleum, Natural Gas and Petrochemical; Aeronautics; Mining; Iron and steel industry; Paper and Cellulose [industry], and Meat' (BNDES, 2017, p. 135).

Brazil's aspirations to become a regional power are positively associated with funding from infrastructure investment projects throughout the continent, as well as the internationalization of Brazilian companies, particularly in South America (BNDES, 2017). Given the prominent role of the state in both processes, they also contain elements of a national development project, after 2003, mainly based on neo-developmentalist or neo-extractivist ideas.[12] However, because of the economic recession (OECD, 2016a), the recent fall of Brazilian FDI (CNI, 2013) in Latin America, coupled with a larger share of FDI associated with speculative capital in tax havens, has weakened the country's aspirations as a regional power.

Recently, reacting to the economic crisis, the Temer government has implemented policies of austerity and cuts in public spending, withdrawing resources from the BNDES.

In summary, the political role of Brazil as a regional power was based on South–South, and regional, co-operation narratives during the 2000s. Such leadership was exercised under a combination of treaties and agreements within Mercosur, within BRICS – always expressed in contradictory terms of co-operation and competition – and in public investments, mainly by way of BNDES loans and the acquisition of shares in Brazilian firms supporting the 'national champions policy' (BNDES, 2010).

Table 2. Infrastructure projects funded by BNDES loans in South America for the export of engineering and other services (2008).

Countries	Amount of loan, US$ million
Argentina	516
Bolivia	152.1
Chile	208
Colombia	678
Cuba	682
Ecuador	511.2
Paraguay	77
Peru	420
Uruguay	29
Venezuela	218.9
Total	2610

Source: Iglesias (2008).

Brazil in Latin America: agrarian dynamics, food security, and agribusiness

Brazil's contradictory or ambiguous role as a regional power is clearly seen in its influence on policy-making in Latin America. Reproducing and reinforcing its own historical contradictions, its leadership has tried to combine public support for the internationalization of large agribusiness firms (increasing the levels of competition in the international market) with encouraging the replication of family farming and food security policies. Such a political trend expresses the ambiguous nature of the Brazilian neo-developmentalist model, and jeopardizes the country's influence as a regional power.

Latin America's role in the world economy has always been related to its abundant natural resources and, specifically, the agro-export model in the case of Brazil. This structural landscape did not change in a significant way after the industrialization period marked by import substitution, as discussed by scholars in the 1960s and 1970s, like Frank (1967), Cardoso (1972), and Marini (1973). In contrast to East Asia, the political organization of newly created industrial elites did not undermine the political power of the agrarian elites (Evans, 1995). A structural economic change such as the one that took place among latecomers in East Asia has never been realized in Latin America (De Janvry, 1981).

Therefore, natural resources are still the major source of wealth and political power for most Latin American countries (Branco, 2013). This obviously has several implications for the modernization of these countries, helping to explain high levels of inequality, low levels of social protection, and a technological dependency on developed capitalist countries.

Moreover, some Latin American countries, such as Brazil and Argentina, became very competitive in agricultural and animal production over the past 30 years. Together with the recent commodity boom, this contributed to agriculture and agribusiness being crucial economic drivers in the region. In terms of agrarian changes, current trends in Latin America are characterized by a deepening dualistic model of access to land and resources (Balestro, 2015). On the one hand, there has been a growing use of natural resources such as land and water by the agribusiness sector and minerals by the mining companies. This process not only involved distribution of wealth, but also had a profound impact on the environment, which, in turn, contributed to an increase in social conflict. On the other hand, it must be recognized that social movements and various other collective actors (co-operatives, associations, etc.) from the countryside managed to create political space, building alternatives from the grassroots initiatives they represent (Petras & Veltmeyer, 2001).

Since the beginning of the 2000s and the end of dictatorships in Latin America (mid-1990s), social actors and civil society organizations have been largely able to resume activities and mobilization (Alvarez, Dagnino, & Escobar, 1999). Particularly in rural areas throughout the region, new social movement shave sprung up, working for indigenous peoples' claims, the restitution of land, access to water and preservation of their identity, as well as providing credit and support to small-scale producers or family farmers, whose claims started to be recognized and taken into account (Wolford, 2010, 2015).

When Latin American centre-left parties took power, several of the proposals and claims made by those organizations became part of government policy, especially in the case of land rights, food security, and social issues.

In rural areas, there was a significant change in policy to support and enrich family farming, which has attracted increasing interest from various quarters. Despite the inequity of its agrarian structure, Brazil was at the forefront of implementing government programmes – though not without harsh criticism – that promote access to land by settlement projects via various purchase mechanisms (Silveira et al., 2016) as well as the expropriation of land for agrarian reform (Sauer, 2017). This is why the Brazilian state's influence is crucial across Latin America, as support is given to large investments in agribusiness alongside public policies that target family farming (Grisa & Schneider, 2015).[13]

In fact, from the 2000s onwards, Brazil developed a particular approach to family farming that relies on credit policies – particularly, an increase in the budget and credit lines of the National Programme Supporting Family Farming, access to technologies and a system of government purchases, which culminated in special legislation for this social category. Brazil is one of a few countries to have established a specific, legal definition of family farming by law (Act 11.326, 2006), which takes into account criteria like land size and the use of family members' labour (Grisa & Schneider, 2015). To a certain extent, this played a role in the transfer of family farming policy to other Latin American countries.

Simultaneously, in a number of countries, public policies were set up to respond to such demands, especially to expand the recognition and legitimacy of family farming in LAC (Schneider, 2014). The increasingly frequent use of the concept of family farming has occurred in several countries and nowadays is widely recognized by scholars and policy-makers (Craviotti, 2014). It is important to remember that Brazil had influence in organizations such as the FAO, led by José Graziano da Silva, while some Brazilian policies for family farming and food security were also a reference for other Latin American countries (RIMISP/IFAD, 2014; Salcedo & Guzmán, 2014).

There is a growing consensus that family farming has acquired a major role in rural development and food security in Central America (Salcedo & Guzmán, 2014; Schneider, 2016). Recently, initiatives in the region, specifically the Family Farming Plan in El Salvador, have also been important in disseminating public policies supporting family farming (Schneider, 2014). The same is happening in countries like Ecuador (Gómez, Le COQ, & Samper, 2014), Peru (Eguren & Pintado, 2015) and, more recently, Colombia (Acevedo-Osorio & Martínez-Collazos, 2016).

Research suggests that family farming could fit into policy goals more effectively by aligning it better with legal frameworks. Family farming is broad social category; consequently, public opinion can readily acknowledge its potential contribution to food supply and security (Schneider, 2014). As a concept, it facilitates the measurement of economic and social indicators concerning its size and contribution to overall production in agri-food systems both domestically and abroad (Grisa & Schneider, 2015).

Relatedly, Brazil pioneered an important regional initiative for rural policies, namely the Family Farming Specialised Network (REAF) among Mercosur member countries. This initiative, begun in 2004, has been crucial for the dissemination of the concept of family farming forged in Brazil in the 1990s (Schneider, 2014).As a driver of regional integration, Brazil played an exceptional role in the creation of the REAF, a successful initiative that not only built a common definition for family farming, but also shared expertise and public policies with this productive sector.

Being a political initiative of the Brazilian government, REAF was included in a broader context of foreign policy reorientation aimed at regional integration and the strengthening of ties with developing countries (Sabourin et al., 2015). The main policy instruments discussed by REAF were subsidized interest rates, phytosanitary measures, technology generation and transfer, price regulation, credit, crop insurance subsidies, and access to land and production inputs.[14] The main purpose was to reduce asymmetries between countries and promote income by facilitating marketing for family production.

Although it is impossible to gauge the exact economic impact of these initiatives so far, the dissemination of the concept and policy transfer of family farming has had positive effects. In particular, the family farming concept has guided the formation of agencies, programmes, and public policies in countries like Argentina, Uruguay, Paraguay, and more recently Nicaragua and El Salvador (Franca & Peracy Sanches, 2015; Patriota, Pierri, Maclennan, & Salles, 2015).

Along with the dissemination of policies on family farming, discussions around food and nutrition security also benefited from considerable contributions and influence on the part of Brazil. This influence was expressed, in part, through the political role played by the regional director of FAO, José Graziano da Silva (2005–2012), who was responsible for the implementation of the Zero Hunger Programme in Brazil (Schneider, 2014), which was deployed as part of the Latin America *Sin Hambre* Programme (Giunta, 2014; Rubio, 2014). Related to food and nutrition security, and specifically to problems of obesity, it has also seen Brazil actively participate within the Pan-American Health Organisation (Souza & Belik, 2012).

In terms of regional influence, especially related to issues of land and agriculture investments, it is important to acknowledge that this is a 'two-way' process. According to BNDES (2014), there is a strong presence of foreign and multinational companies in Brazilian agribusiness, including Latin American companies. For example, through mergers and acquisitions, corporate control of the Brazilian sugarcane industry has become dominated by foreign capital. Of the five largest companies, two are foreign multinationals (McKay, Sauer, Richardson, & Herre, 2015).[15]

As can be seen in Table 3, among the 15 largest agribusiness companies doing business in Brazil, only 5 are foreign multinationals. In comparison with the 15 largest companies in the electronics industry, there is a substantial difference – five of them are Brazilian companies, while the others are multinationals. Together with mining companies, civil engineering, and infrastructure building contractors, agribusiness is by far the most relevant sector for large national economic groups.

Agribusiness companies also account for 21% of the total BNDES (2010) loans to Brazilian multinationals, which amounted to US$27.9 million in 2010. According to Mackey (2015, p. 7), 'the large size of these firms means that the activities of one or a handful of organizations may nonetheless, have an outsized financial impact on agro-industrial restructuring in Latin America'. This is best demonstrated by the aforementioned JBS group, which posted a revenue of more than US$39 billion in 2013 and has since become the largest private firm in Brazil and the biggest meat processor in the world.[16]

There is evidence that agribusiness groups are becoming more heterogeneous as well, including firms and co-operatives that specialize in grains (Mackey, 2015). At the same time, especially

Table 3. Major agribusiness companies in Brazil by revenue and country of origin.

Companies	Country of origin	Revenue (US$ million)
JBS	Brazil	33,464
Ambev	Brazil and Belgium	10,578
Bunge Alimentos	Holland	9468
BRF	Brazil	8057
Cargill	USA	7264
Coopersucar	Brazil	5829
Coamo	Brazil	2265
Tereos Internacional	France	2233
Nidera Sementes	Argentina	1616
Cia. Vale	Brazil	1281
Biosev	France	1254
Lar	Brazil	839
Cocamar	Brazil	762
Odebrecht Agroindustrial	Brazil	745
Santa Terezinha Participações	Brazil	574

Source: Valor 1000. Available at http://www.valor.com.br/valor1000/2015/ranking1000maiores.

between 2006 and 2010, there was very substantial foreign investment by oil companies such as Shell and British Petroleum and 'food' companies like Bunge and Cargill in the Brazilian sugarcane sector (see McKay et al., 2015; Sauer, Pietrafesa, & Pietrafesa, 2017) (Table 4).

Agriculture and livestock husbandry account for a significant share of BNDES loans and investments (BNDES, 2017). Besides BNDES, governmental financial support to agribusiness remains important, since its annual budget of around US$45.6 billion was used for credits and loans in 2015/2016. This sector has been considered important in terms of co-operation projects with other Latin American countries, which are mostly carried out by EMBRAPA as part of its technical co-operation programme. However, Brazilian agribusiness has played a significant international role in agrarian transformation (land use, for instance) in Latin America (in terms of expanding soy cultivation) and, especially, in neighbouring countries (Wesz, 2011).

In summary, Brazil's role as a leading regional power has been based on supporting large agribusiness firms to induce or increase competition in the foreign market. Reproducing internal social and political contradictions, its public financial support has prioritized large firms, not only increasing competition but also favouring a concentration of resources in a tiny number of companies, of which JBS is a prime example, with no particular gains for a national or regional development process.

Further dimensions of Brazil as a regional power

With the Lula administration, the main foreign policy-making group emphasized on Brazil's autonomy. It stood for a more self-directed and active projection for the country in the international arena (Saraiva, 2010). Lula's foreign policy towards South America was driven by a strong belief in the country's destiny as a global power. Undoubtedly, the strengthening of co-operation between Latin America and Africa, concerning the transfer of policy experiences that had proved successful domestically, was a clear indicator of the government's soft power strategy.

The Lula government's wager on South–South relations allowed for Brazil to play a more prominent role in bilateral and multilateral co-operation projects, especially in Latin America. According to the ABC (2014), under the management of the Ministry of Foreign Affairs, 'international technical cooperation constitutes an important development tool, helping the countries to promote structural changes in social and economic fields, including state management, through institutional strengthening actions'.

Latin America accounted for 68% of federal government disbursement for international co-operation, followed by Africa with 23% in 2014 (see Figure 6). Such disbursements did not do justice to the frequent argument in favour of co-operation as solidarity with the African continent, in respect of which Brazil owes a historical debt due to slavery (ABC, 2014).

Within Latin America, most of the co-operation projects took place in low-income countries. Out of 57 such projects, 48 were carried out in co-operation with poorer countries from South and

Table 4. BNDES loans to Brazilian multinationals in the food industry, 2008–2010.

Companies	Amount of loan (US$ million)	Number of branches in Latin America
JBS	4.274	38
BR foods	924	37
Marfrig	629	18
Minerva	70	5
Total	5897	98

Source: Mackey (2015, p. 7).

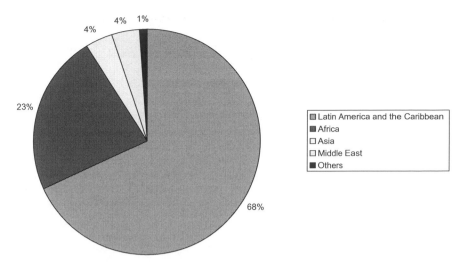

Figure 6. Distribution of Brazilian federal government funding for international co-operation per region. Source: ABC (2014).

Central America (Table 5). Potentially, this could translate into Brazilian policy ideas having greater regional sway.

Between 2005 and 2009, around 76% of total Brazilian humanitarian assistance went to Latin American countries. However, when the figures concerning expenditures by type of co-operation are analysed, contributions to international organizations such as the UN are almost double the amount given to humanitarian co-operation (US$312 million versus US$161 million) (see Table 6), and there are no terms of comparison with public resources for investments and loans. Besides, the amounts are small in comparison to China's annual aid budget, which averagesUS$5 billion.[17] It is worth noting that China is the 10th largest donor in the world.

Brazilian technical co-operation operates in several fields, including health, agriculture, vocational education (scholarship), water resources, public administration, and energy. However, the largest

Table 5. Co-operation projects in Latin America.

Countries	Number of co-operation projects related to agriculture and rural development	World Bank classification according to income level
Argentina	1	High income (non-OECD)
Bolivia	9	Lower middle income
Colombia	6	Upper middle income
Ecuador	6	Upper middle income
Guyana	4	Lower middle income
Paraguay	4	Upper middle income
Uruguay	2	High income (non-OECD)
Venezuela	6	High income (non-OECD)
Costa Rica	2	Upper middle income
Nicaragua	4	Lower middle income
Panama	5	Upper middle income
Jamaica	2	Upper middle income
Haiti	3	Low income
Cuba	3	Upper middle income
Total	57	

Source: ABC (2014).

Table 6. Brazilian expenditure by type of co-operation, 2010 (US$ million).

Type of co-operation	2010
Contributions made to international organizations	312
Humanitarian co-operation	161
Technical co-operation	58
Educational co-operation	36
Scientific and technological co-operation	24
Total	591

Source: ABC (2014).

budget (22% of the total amount) between 2003 and 2010 was allocated to agricultural co-operation (Renzio, de, Gomes, Fonseca, & Niv, 2013). In 2013, there were 47 co-operation projects involving partnerships with the EMBRAPA, with 39 taking place in African countries, the most important being the Pro-Savana programme in Mozambique (Renzio et al., 2013; Scoones, Amanor, Favareto, & Qi, 2016), which was designed to take the Green Revolution model to Africa.

According to Costa Leite and others (2014), there seems to be a consensus that Brazil is going through a transitional moment regarding its role in international co-operation. The creative experience of 'inclusive growth' in the middle of the global crisis was rather alluring to international organizations, despite the social indicators falling in most countries. Prior to 2016, this narrative was important in establishing Brazilian policies and programmes as 'best practices' to be shared with other developing countries (Costa Leite et al., 2014).

Aside from the meagre financial resources allocated to international co-operation, Brazil still faces some major obstacles to South–South co-operation (Costa Leite et al., 2014). That there is no legal framework consistent with the country's role as a co-operation provider is one of them. Another is the absence of a strong and autonomous co-ordinating agency, which has led to fragmented strategies without a long-term and consistent international co-operation policy. The fact that Brazilian international co-operation depended too much on the individual initiatives of bureaucrats and politicians caused its rapid erosion after the radical change in government in 2016. There is also the need for funding channels and adequate information regarding Brazilian co-operation for different stakeholders to engage in South–South co-operation.

The recent trend (after 2016) in Brazilian foreign policy away from a more autonomous position towards a greater alignment with North American and European foreign policies reveals the problems concerning the level of statehood derived from soft power initiatives; in other words, the lack of continuity in South–South co-operation, and aspirations to be an emerging global power, show that there were not enough efforts to build state capability to design and implement long-term policies autonomously from changes in government. As with the global powers, foreign policy should not be subject to the whims of the current political administration.

Conclusion

The process of Brazil's recent rise as a regional power in Latin America is ambiguous and unstable. Specifically, its impact on other Latin American countries is uncertain. On the one hand, there are positive lessons and good examples from Brazil in respect of social and rural development policies, as well as South–South international co-operation. On the other hand, the failure to overcome the neo-extractivist model alongside deindustrialization and rentierism has severely undermined the trajectory of the country as a global player in the world economic order. Recent backlashes in Brazilian foreign policy mirror this new reality. Since 2016, the country has lost a great deal of its autonomy in the geopolitical game.

The rural and social policies transferred from Brazil to other countries are not restricted to Latin America. Co-operation with Africa has become more relevant over the last decade. African nations have attracted many visits by Brazilian authorities, including the president. Such visits resulted in the opening of new embassies in Africa. Similar to Latin America, agrarian issues also have had an impact, as has been the case with the famous Pro-Savana project in Mozambique (Cabral, Favareto, Mukwereza, & Amanor, 2016).

The shaky stance of Brazil as a regional power is closely tied to the fact that the narrative of 'developmentalism' – and the national development project – has run out of steam and has been eroded as a result. One major reason for this is the lack of complementarity between industrial policies and macroeconomic policies conducive to economic growth. Notwithstanding the huge fiscal incentives for manufacturing, the situation among domestic manufacturing firms has deteriorated due to an overvalued Brazilian Real and skyrocketing interest rates (among the highest in the world). In addition, the increase in real wages has not been followed by an increase in productivity. Commodity-boom dependency and deindustrialization are now taking their toll, with severely negative consequences for the country's economic growth as well as Brazilian FDI. Brazil's strategy – or the inadequacy thereof – has been reflected in the tensions and contradictions within coalition government led by the Workers' Party.

Brazil is now going through one of the deepest political and economic crises in its history as a republic, and a neoliberal direction looms large. Such a course, concerning the role of the state in economic and social development in Brazil, and the end of the commodity boom, threatens the country's position as a regional power. The country is much less engaged in South–South co-operation. It has abandoned any political effort to improve regional leadership, and there are fewer financial resources for co-operation projects with other Latin American countries. Economically, there is a stronger dependence on multinationals – including agribusiness companies – buying Brazilian firms. The degree of resilience the legacy of the golden decade of Brazilian leadership in Latin America will show, and the extent of co-operation with other members of the BRICS, remains to be seen.

Notes

1. Jenkins and Barbosa (2012, p. 68), however, in disagreeing with the 'deindustrialisation' thesis state that 'even though Brazil is not deindustrialising, its manufacturing industry needs to adapt its structure to a new context of more dynamic internal growth and increasing Chinese competition'.
2. China overtook the United States as the top export destination in 2013 by purchasing 14% of South American exports, compared to the 12% purchased by the United States. The total of LAC exports to China grew to US$112 billion in 2013 (a record 2.0% of regional GDP), though the region still had a trade deficit of 0.5% of its GDP with the Asian country that year. In contrast, China bought only 2% of exports from Mexico, Central America, and the Caribbean, where the US still had a dominant market share of 69% (Brent, 2015).
3. The Complementary Law 87 – commonly known as the Kandir Act, named after Cardoso's Ministry of Finance in 1996 – exempts taxes for exporting unprocessed or semi-processed primary products. All agricultural commodities and minerals are exported virtually tax free (Sauer & Mészáros, 2017), leading to estimated losses of more than R$718 billion (US$228.2 billion at the 2017 currency rate) since 1996.
4. From 2009 to 2013, 69% of Chinese mergers and acquisitions went to oil and gas investments (McKay et al., 2016); more FDI focused on energy, infrastructure, and communications and less on land and agriculture in LAC. However, China's investments changed dramatically after 2013 because of one major project, the construction of the Nicaragua Canal (Brent, 2015).

5. Related to incomes, the minimum wage in Brazil increased by 81% between 2003 and 2010, after taking inflation into account. As for the development of the formal economy, employment grew by more than 15 million during the same period, and the unemployment rate was only 6.5% in 2013 (OECD, 2016b), which was considered a decisive factor in poverty reduction; however, the situation changed drastically after 2014, with unemployment rates increasing up to 13.7% in 2017 (Silveira & Cavallini, 2017).

6. There were other initiatives after 2003, like strengthening of – or, at least, the resumption of negotiations within – Mercosur, and the creation of the South American Defense Council and of the Union of South American Nations (UNASUR). According to Kellogg (2007, p. 189), 'The UNASUR, while a challenge to US hegemony in the region, is completely embedded in a very familiar logic of capital accumulation and corporate rule.'

7. According to Cervo and Lessa (2014), the decline of Brazil as an emerging power under the Dilma Rousseff government after 2011 had not shaken South–South coalitions and strategic partnerships in the region, but it did result in a stronger role for BRICS, especially for its two larger countries, China and Russia.

8. According to a 2016 report, Globo Holding's marketable securities accounted for 33.3% of the total current assets of the holding company in 2015. Total current assets from Globo Holding were 10.3 billion reals in 2015 and marketable securities were 3.4 billion reals. In 2016, marketable securities accounted for 40% (2.6 billion reals) of the total current assets (6.5 billion reals). Report available at https://globoir.globo.com/ (Accessed 30 August 2017).

9. According to Moraes (2017), the BNDES loaned around R$12.8 billion (US$3.87 billion in 2017 currency rates) to the JBS group between 2002 and 2013, allowing it to purchase of several companies in the meat sector, including refrigeration facilities in Argentina, Uruguay, Paraguay, the United States, Canada, and Australia. The JBS group raised its revenues from R$4 billion (US$1.2 billion) in 2006 to R$170 billion (US$51.4 billion) in 2016, transforming it in the world largest meat processor.

10. Seeking to respond to criticisms of waste of public resources by the PT administrations, the new president of BNDES, appointed by the current government, released a report – a so-called Livro verde or Green Book, probably alluding to 'green card' or sustainability – on financial operations between 2001 and 2016. Defending these operations, the report concludes that they generated significant profit for the bank (see BNDES, 2017).

11. After a meeting in Fortaleza, in 2014, the five BRICS heads of state met with 11 South American presidents in order to extend regional co-operation via the BRICS Bank, especially for infrastructure projects (Cervo & Lessa, 2014). Along the same path, China accounted for over half of all new FDI (known as greenfield FDI, or GFDI) in LAC in 2013, especially because of the project of the Nicaragua Canal (Brent, 2015; McKay et al., 2016).

12. As mentioned above, such perspective changed radically in 2016 with the Temer government and its neoliberal economic programme of austerity and the reform of labour and social security policies and laws. This neo-extractivism model already unstable after 2011 (see Figure 5), deepening with the economic crisis after 2014 and the 'need' for cutting down on public expenditure and investment (Sauer & Mészarós, 2017).

13. The governmental support to family farming is clearly seen in its increasing budget, which rose from R$2.3billion (US$640 million), in 2003, to R$28.9 billion (US$7.2 billion) in 2016. However, this family farming's budget represented only 15.5% of the R$187.7 billion (US$ 46.6 billion) available for agribusiness in 2016, showing the disparity in governmental allocation of resources (Sauer, 2015, 2017).

14. 'This is a new process that started in the mid-1990s in Brazil, which evolved and, from the 2000s onward, was disseminated to other countries in the region' (Craviotti, 2014, p. 47).

15. Even with such an increase in foreign investments, the sugar/ethanol sector championed the BNDES loans, since it received around R$55 billion (US$17.5 billion in 2017 currency exchange rates) between 2004 and 2010, allowing for the implementation of more than 120 projects, which practically doubled the productive capacity of the sector (BNDES, 2017), as part of the government's national plan for energy (see Sauer et al., 2017).

16. According to Mackey (2015, p. 3), 'a small number of vertically-integrated groups drive uneven and multi-scalar processes that principally operate in sectors that are downstream of livestock production in subnational regions of the Southern Cone'. Mackey's (2015, p. 3) main argument is:

that agro-industrial production by firms from Brazil in Latin America is an uneven and multi-scalar process by a small number of firms that specialize in downstream sectors and livestock and grains commodity network rather than a Brazilian agri-food globalization across Latin America.

17. See China's Second White Paper on Foreign Aid Signals Key Shift in Aid Delivery Strategy at http://asiafoundation.org/in-asia/2014/07/23/chinas-second-white-paper-on-foreign-aid-signals-key-shift-in-aid-delivery-strategy/. Accessed 9 January 2016.

Disclosure statement

No potential conflict of interest was reported by the authors.

Funding

The authors would like to thank to CNPq for scholar grants (Sergio Schneider, PQ scholarship no. 306165/2016-3; and Sérgio Sauer – PQ scholarship no. 303677/2016-3) and also thank CAPES for a visiting scholar grant (no. 10912/13-4) to Moisés Balestro.

References

ABC-Agência Brasileira de Cooperação. (2014). *Brazilian cooperation for international development 2010.* (R. Baumann, Ed.). Brasília: Ipea-ABC.

Acevedo-Osorio, A., & Martínez-Collazos, J. (2016). *La agricultura familiar en Colombia. Estudios de caso desde la multifuncionalidad y su aporte a la paz.* Bogotá: Ediciones de la Universidad Cooperativa de Colombia.

Alvarez, S., Dagnino, E., & Escobar, A. (Eds.). (1999). *Cultures of politics/politics of cultures: Re-visioning Latin American social movements.* Boulder, CO: Westview Press.

Balestro, M. (2015). *Financialization in agribusiness: Some notes from the Brazilian case* (Bicas Working Paper). Brasília: University of Brasília.

Baletti, B. (2014). Saving the Amazon? Sustainable soy and the new extractivism. *Environment and Planning A, 46,* 5–25.

BNDES. (2014, December). *Perspectivas do investimento 2015–2018 e panoramas setoriais.* Rio de Janeiro: APE/DEPEQ, BNDES. Retrieved from https://web.bndes.gov.br/bib/jspui/bitstream/1408/2842/7/Perspectivas%20do%20investimento%202015-2018%20e%20panoramas%20setoriais_atualizado_BD.pdf

BNDES. (2017, July). *Livro verde: Nossa história como ela é.* Rio de janeiro: BNDES (Preliminary draft). Retrieved from https://web.bndes.gov.br/bib/jspui/bitstream/1408/12697/1/Livro%20Verde.pdf

BNDES-Banco Nacional de Desenvolvimento Econômico e Social. (2010). *Relatório Anual – 2010*. Rio de Janeiro: BNDES. Retrieved from http://www.bndes.gov.br

Boito, A., & Berringer, T. (2014). Social classes, neodevelopmentalism, and Brazilian foreign policy under Presidents Lula and Dilma. *Latin American Perspectives*, *41*, 94–109.

Branco, R. S. (2013). Raul Prebisch e o desenvolvimento econômico brasileiro recente liderado por commodities. *Sociais e Humanas*, *26*(1), 197–216.

Brent, Z. (2015, June 5–6). *The rise of BRICS and MICs: China in Latin America*. International conference on land grabbing, conflict and agrarian-environmental transformations: Perspectives from East and Southeast Asia, Chiang Mai University (Working paper).

Bresser-Pereira, L. C. (2015). Reflecting on new developmentalism and classical developmentalism. *Review of Keynesian Economics*, *4*(3), 331–352.

Burchardt, H.-J., & Dietz, K. (2014). (Neo-)extractivism: A new challenge for development theory from Latin America. *Third World Quarterly*, *35*(3), 468–486.

Cabral, L., Favareto, A., Mukwereza, L., & Amanor, K. (2016). Brazil's agricultural politics in Africa: More food International and the disputed meanings of 'family farming'. *World Development*, *81*(May), 47–60.

Cano, W. (2012). A desindustrialização no Brasil. *Economia E Sociedade*, *21*, 831–851.

Cardoso, F. H. (1972). Dependency and development in Latin America. *New Left Review*, *74*, 83–95.

Carpintero, Ó., Murray, I., & Bellver, J. (2016). The new scramble for Africa: BRICS strategies in a multipolar world. *Research in Political Economy*, *30B*, 191–226.

Cervo, A. L., & Lessa, A. C. (2014, July/December). O declínio: inserção internacional do Brasil (2011–2014). *Revista Brasileira de Política Internacional*, *57*(2), 133–151.

Comisión Económica para América Latina y el Caribe (CEPAL). (2016). *Panorama de la Inserción Internacional de América Latina y el Caribe, 2016 (LC/G.2697-P)*. Santiago: CEPAL.

Conceição, J. C. P. R., & Conceição, P. H. Z. (2014). *Agricultura: Evolução e importância para a balança comercial brasileira* (Texto para Discussão no. 1944). Brasília: Instituto de Pesquisa Econômica Aplicada.

Confederação Nacional da Indústria (CNI). (2013). *Os investimentos brasileiros no exterior: relatório 2013*. Brasília: Author.

Costa Leite, I., Suyama, B., Trajber Waisbich, L., Pomeroy, M., Constantine, J., Navas-Alemán, L., … Younis, M. (2014). *Brazil's engagement in international development cooperation: The state of the debate* (IDS Evidence Report 59).

Craviotti, C. (2014). *Agricultura familiar en Latinoamérica*. Buenos Aires: Editorial CICCUS.

Curado, M. (2015). China rising: Threats and opportunities for Brazil. *Latin American Perspectives*, *42*(205), 88–104.

De Janvry, A. (1981). *The agrarian question and reformism in Latin America*. Baltimore, MD: Johns Hopkins University Press.

De la Torre, A., Didier, T., Ize, A., Lederman, D., & Schmukler, S. (2015). *Latin America and the rising south: Changing world, changing priorities*. Washington, DC: World Bank Group.

Doctor, M. (2015). Assessing the changing roles of the Brazilian Development Bank. *Bulletin of Latin American Research*, *34*(2), 197–213.

ECLAC. (2014). *Panorama social de América Latina 2014*. Santiago: United Nations.

ECLAC – The Economic Commission for Latin America. (2013). *Chinese foreign direct investment in Latin America and the Caribbean: China-Latin America cross-council task force*. Santiago: United Nations.

Eguren, F., & Pintado, M. (2015). *Contribución de la agricultura familiar al sector agropecuario en el Perú*. Lima: Centro Peruano de Estudios Sociales (CEPES).

Evans, P. (1995). *Embedded autonomy: States and industrial transformation*. Princeton, NJ: Princeton University Press.

Food and Agriculture Organisation (FAO), International Fund for Agricultural Development (IFAD) & World Food Programme (WFP). (2013). *The state of food insecurity in the world 2013. The multiple dimensions of food security*. Rome: FAO.

Franca, C. G., & Peracy Sanches, A. (2015). Public policies for the strengthening of family farming in the Global South. *Policy in Focus*, *12*(4), 11–15.

Frank, A. G. (1967). *Capitalism and underdevelopment in Latin America: Historical studies of Chile and Brazil*. New York, NY: Monthy Review Press.

Freitas, G. (2011, October 24). A última fronteira do capital especulativo. *Valor Econômico*.

Giunta, I. (2014). Food sovereignty in Ecuador: Peasant struggles and the challenge of institutionalization. *The Journal of Peasant Studies, 41*(6), 1201–1224.

Gómez, I., Le COQ, J. F., & Samper, M. (2014). *Las agriculturas familiares en Centroamérica: procesos y perspectivas*. San Salvador: PRISMA.

Grisa, C., & Schneider, S. (2015). *Políticas públicas de desenvolvimento rural no Brasil*. Porto Alegre: UFGRS.

Iglesias, R. (2008). Some elements to characterize Brazilian interest in infrastructure integration in South America. *Integration & Trade Journal, 12*(28), 149–178.

Instituto de Pesquisa Econômica Aplicada (IPEA). (2010). *Brazilian cooperation for international development*. Brasília: Ipea and ABC.

Jenkins, R. (2015, November). Is Chinese competition causing deindustrialization in Brazil? *Latin American Perspectives, 42*(6), 42–63.

Jenkins, R., & Barbosa, A. F. (2012). Fear for manufacturing? China and the future of industry in Brazil and Latin America. *The China Quarterly, 209*, 59–81.

Kellogg, P. (2007). Regional integration in Latin America: Dawn of an alternative to neoliberalism? *New Political Science, 29*(2), 187–209.

Lazzarini, S., & Musacchio, A. (2014). *Reinventing state capitalism: Leviathan in business, Brazil and beyond*. Cambridge, MA: Harvard University Press.

Mackey, L. (2015). *Beyond the fence of the Brazilian farm: New evidence on Brazil-based agroindustry in Latin America* (Bicas Working Paper). São Paulo: BRICS Initiative for Critical Agrarian Studies.

Marini, R. M. (1973). *Dialéctica de la dependencia*. México, DF: Ediciones Era.

McKay, B., Alonso-Fradejas, A., Brent, Z. W., Sauer, S., & Xu, Y. (2016). China and Latin America: Towards a new consensus of resource control? *Third World Thematics: A TWQ Journal, 1*(5), 592–611.

McKay, B., Sauer, S., Richardson, B., & Herre, R. (2015). The political economy of sugarcane flexing: Initial insights from Brazil, Southern Africa and Cambodia. *The Journal of Peasant Studies, 43*, 1–29.

MDIC – Desenvolvimento, Indústria e Comércio. (2015). *Estatísticas de comércio exterior*. Brasília. Retrieved from www.desenvolvimento.gov.br//sitio/interna/index.php?area=5

Moraes, M. d. (2017, May 23). Saiba como a JBS sugou o BNDES para expandir seus negócios. *Jornal Estado de Minas (Em.com.br)*. Retrieved from http://www.em.com.br/app/noticia/economia/2017/05/23/internas_economia,871042/saiba-como-a-jbs-sugou-o-bndes-para-expandir-seus-negocios.shtml

Organisation for Economic Co-operation and Development (OECD). (2016a). *Latin American economic outlook 2016: Towards a new partnership with China*. Paris: Author.

Organisation for Economic Co-operation and Development (OECD). (2016b). *Labour force statistics 2015: Brazil*. Paris: Author. Retrieved from https://data.oecd.org/unemp/unemployment-rate.htm

Patriota, T. C., Pierri, F. M., Maclennan, M., & Salles, M. (2015). The growing recognition of family farming as part of the solution of sustainable development: Evidence from recent policy shifts. *Policy in Focus, 12*(4), 4–7.

Petras, J., & Veltmeyer, H. (2001). Are Latin American peasant movements still a force for change? Some new paradigms revisited. *The Journal of Peasant Studies, 28*(2), 83–118.

Ray, R., & Gallagher, K. (2017). *China-Latin America Economic Bulletin* (Discussion Paper 2015-9). Global Economic Governance Initiative. Retrieved from https://www.bu.edu/pardeeschool/files/2014/11/Economic-Bulletin.16-17-Bulletin.Draft.pdf

Renzio, P., de, Gomes, G. Z., Fonseca, J. M. E. M., & Niv, A. (2013). *O Brasil e a cooperação Sul-Sul: Como responder aos desafios correntes* (BPC Policy Brief, Vol. 3, No. 55). Rio de Janeiro: BRICS Policy Center.

RIMISP-Latin-American Centre for Rural Development & IFAD-International Fund for Agricultural Development. (2014, October). *Family farming in Latin America. A new comparative analysis* (Synthesis Report). Rome: IFAD.

Rodrik, D. (2016). Premature deindustrialisation. *Journal of Economic Growth, 21*(1), 1–33.

Rubio, B. (2014). *El dominio del hambre. Crisis de hegemonía y alimentos*. Chapingo: Universidad Autónoma de Chapingo, Colegio de Posgraduados, Juan Pablos.

Sabourin, E., Samper, M., & Sotomayor, O. (2015). *Políticas publicas y agriculturas familiares en América Latina y el Caribe: Nuevas perspectivas*. San José: IICA.

Safransky, S., & Wolford, W. (2011, April 6–8). *Contemporary land grabs and their alternatives in the Americas*. International conference on Global Land Grabbing, University of Sussex. Retrieved from www.future-agricultures.org/index

Salama, P. (2013, September). *Les économies émergentes, le plongeon?* (FMSH-WP-2013-42).

Salcedo, S., & Guzmán, L. (Eds.). (2014). *Agricultura familiar en América Latina y el Caribe: Recomendaciones de política.* Santiago: FAO.

Saraiva, M. G. (2010). Brazilian foreign policy towards South America during the Lula administration: Caught between South America and Mercosur. *Revista Brasileira de Política Internacional, 53,* 151–168.

Sauer, S. (2015). *Brazil and Latin American agrarian transformations: A preliminary survey (oral presentation).* Bicas conference on 'Rural transformations and food systems: The BRICS and agrarian change in the global South', Institute for Poverty, Land and Agrarian Studies, Cape Town.

Sauer, S. (2017). Rural Brazil during the Lula administrations: Agreements with agribusiness and disputes in agrarian policies. *Latin American Perspectives.* Online first, 1–19. doi:10.1177/0094582X16685176

Sauer, S., & Mészarós, G. (2017). The political economy of land struggle in Brazil under workers' party governments. *The Journal of Agrarian Change, 17,* 397–414.

Sauer, S., Pietrafesa, J. P., & Pietrafesa, P. A. (2017). Climate change and agrofuels: Brazilian ethanol and the Cerrado biome. In M. J. Angelo & A. Du Plessis (Eds.), *Research handbook on climate change and agricultural law* (pp. 331–366). Cheltenham: E. Elgar publishing.

Schneider, B. R. (2009). Hierarchical market economies and varieties of capitalism in Latin America. *Journal of Latin American Studies, 41*(3), 553–575.

Schneider, B. R. (2012). Contrasting capitalisms: Latin America in contrasting perspectives. In J. Santiso & J. Johnson-Dayton (Eds.), *The Oxford handbook of Latin American political economy* (pp. 381–402). Oxford: Oxford University Press.

Schneider, B. R. (2013). *Hierarchical capitalism in Latin America: Business, labor, and the challenges of equitable development.* Cambridge: Cambridge University Press.

Schneider, S. (2014). *Family farming in Latin America and the Caribbean: Deep roots.* Roma: FAO.

Schneider, S. (2016). *Family farming in Latin America and the Caribbean: Looking for new paths of rural development and food security* (International Policy Centre for Inclusive Growth (IPC – IG) UNDP, Working Paper No. 137).

Scoones, I., Amanor, K., Favareto, A., & Qi, G. (2016). A new politics of development cooperation? Chinese and Brazilian engagements in African agriculture. *World Development, 81,* 1–12.

Silveira, D., & Cavallini, M. (2017, March 28). Desemprego fica em 13,7% no 1° trimestre de 2017 e atinge 14,2 milhões. *Portal G1.* Retrieved from http://g1.globo.com/economia/noticia/desemprego-fica-em-137-no-1-trimestre-de-2017.ghtml

Silveira, F. G., Arruda, P., Vieira, I., Battestin, S., Campos, A. E., & Silva, W. (2016). *Public policies for rural development and combating poverty in rural areas.* Brasília: UNDP.

Souza, L. R., & Belik, W. (2012). O planejamento da política de alimentação: uma análise a partir dos casos do México, Brasil e Peru. *Revista Segurança Alimentar e Nutricional, Campinas, 19*(2), 111–129.

Spar, D. L. (2001). National policies and domestic politics. In A. Rugman & T. L. Brewer (Eds.), *The Oxford handbook of international business* (pp. 206–231). Oxford: Oxford University Press.

UNCTADSTAT – United Nations Conference on Trade and Development. (2017). *Datacenter – Foreign direct investment: Inward and outward flows and stock, annual, 1970-2016.* Geneva: UNCTAD. Retrieved from http://unctadstat.unctad.org/wds/TableViewer/tableView.aspx?ReportId=96740

Valor Econômico. (2012, May 3). *Interview with Luciano Coutinho.* Rio de Janeiro: President of the BNDES.

Veltmeyer, H. (2017, April 24–26). *The contemporary dynamics of extractive capital(ism) in Latin America.* International Colloquium the Future of Food and Challenges for Agriculture in the 21st Century. Europa Congress Palace. Vitoria Gasteiz, Álava, Basque Country.

Veltmeyer, H., & Petras, J. (Eds.). (2014). *The new extractivism: A post-neoliberal development model or imperialism of the twenty-first century?* London: Zed Books.

Vergara-Camus, L., & Kay, C. (2017). The agrarian political economy of left-wing governments in Latin America: Agribusiness, peasants, and the limits of neo-developmentalism. *The Journal of Agrarian Change, 17,* 415–437.

Wesz, V. J., Jr. (2011). *Dinâmicas e estratégias das agroindústrias de soja no Brasil.* Rio de Janeiro: E-papers.

Wesz, V. J., Jr. (2016). Strategies and hybrid dynamics of soy transnational companies in the Southern Cone. *The Journal of Peasant Studies.* doi:10.1080/03066150.2015.1129496

Wilkinson, J., Wesz, V. J., Jr., & Lopane, A. R. M. (2016). Brazil and China: The agribusiness connection in the Southern Cone context. *Third World Thematics: A TWQ Journal, 1*(5), 726–745.

Wolford, W. (2010). Participatory action by default: Land reform, social movements, and the state in Northeastern Brazil. *The Journal of Peasant Studies, 37*(1), 91–109.

Wolford, W. (2015). Rethinking the revolution: Latin American social movements and the state in the 21st century. In V. Bennett & J. Rubin (Eds.), *Enduring reforms: Progressive activism and visions of change in Latin America's democracies*. University Park: Pennsylvania State University Press.

Agrarian trajectories in Argentina and Brazil: *multilatin* seed firms and the South American soybean chain

Clara Craviotti

ABSTRACT

Since the turn of the century, Argentinian seed firms have been internationalizing their operations, focusing on neighbouring countries, specially Brazil. A 'flex crop' such as soybean has constituted a central focus for their investment. This article analyses investment opportunities and different intellectual property rights as key drivers of internationalization, and examines the ability of firms to develop networks that are both 'inward' and 'outward' in their orientation, as well as the tensions involved. The analysis points to the emergence of South-South flows of capital that aim to strengthen their position within key components of agri-food chains, and the formation of transnational elites grounded in global circuits of accumulation.

Introduction

In recent years, the role of agriculture in modern economies has generated significant interest from academics, governments and members of the public. After several years in which debates on the Agrarian Question were overshadowed by other issues, they are now firmly back on the agenda, because of the abrupt increase in the prices of staple foods in 2007/2008, and increased levels of foreign investment in land and farming enterprises in different countries, among other factors.

These phenomena emerge from tendencies that are profoundly reshaping contemporary agrifood systems – albeit with different nuances in each country and diverse impacts on their agricultural sectors. Common patterns include: the growing integration of agribusiness activities into global value chains; high levels of concentration, particularly in the 'upstream' and 'downstream' components of agricultural value chains; the rapid diffusion of modern biotechnology; and the emergence of new forms of governance that involve the strengthening of private forms of regulation. These developments are often seen as key features of an agrifood regime that has been characterized as 'corporate' in character (McMichael, 2009) – which is not to say that it is devoid of internal tensions and contradictions.

Among recent trends within global agriculture is a growing demand for animal protein within emerging economies, as well as expanded production of biofuels. These mean that countries that are net exporters of these products – and especially of *flex crops*[1] – are strategically well-placed to expand their share of production and trade. In some cases, expansion involves a fundamental shift in development policy, away from previous industrialization and import-substitution strategies, towards what some authors conceptualize as the *reprimarization* of economies (Bastian & Soihet, 2012), or from a political ecology perspective, as *neo-extractivism* (Gudynas, 2012).[2]

The expansion of a flex crop such as soybean is often based on a distinctive configuration of elements: natural resources such as land, a specific technological package comprised of transgenic seeds, no-till methods and an herbicide (glyphosate), as well as a restructuring of units of production. Organizational innovations, such as the emergence of *multilatin* firms in the initial states of agri-food chains, are of particular interest, since they put into question older notions that capital investments always flow from North to South.[3]

Taking this general framework as a starting point, this article seeks to illuminate how the emergence of internationalized firms in countries such as Argentina and Brazil is linked to their location within a changing global division of labour. It considers the factors that have enabled Argentine seed firms to expand into key components of the Brazilian soybean value chain, accompanying – and fostering – the increased importance of this crop within the economies of both countries.

Seed companies are rarely considered when it comes to relationships amongst countries in the Southern Cone, despite the fact that plant breeding influences both the diversity of food available within the agri-food system as a whole and the environmental impacts of different crops (Bonny, 2014). The seed industry also plays a fundamental role in shaping farming practices, through the control of nature, homogenization and standardization that it involves (Hubert, Goulet, Tallon, & Huguenin, 2013). In strictly economic terms, the seed industry represents a highly dynamic sector both in Brazil and Argentina, which respectively represent the fourth and ninth largest markets for seed worldwide (Argentinian Seed Association, 2013).

The aim here is to analyse the development of *multilatin* firms in the seed production component of the soybean chain, paying attention to the rationale behind their expansion in recent years, the business models that they have adopted, and their relationships with other actors in the seed industry. Regarding the latter, the study shows that while *multilatin* firms are subordinate to biotech companies, this is not necessarily the case when considering their internal networks. This is a relatively unexplored topic, since the dominant position of large multinational corporations in the seed industry, especially in the development of transgenic traits has led to neglect the role of breeding companies, especially those originating in developing countries. Research combined in-depth personal interviews carried out in 2014 with key informants, including public sector officials, company representatives and technicians, analysis of articles published in Argentinian newspapers and magazines during 2005–2014, and collection and analysis of relevant documents and available statistical data.

The article is organized in the following sections. The second section analyses the place of the MERCOSUR[4] grouping of countries within global patterns of soybean production, and the factors that account for the growing importance of this crop. The third section of the article provides some historical background on the internationalization of Argentinian firms[5], and the fourth sets out an analysis of dynamic local actors in the seed sector. Finally, some concluding remarks are presented.

Brazil and Argentina as cornerstones of the MERCOSUR soybean sector

The MERCOSUR grouping of countries is prominent in global rankings of food exporting countries, particularly in relation to soybean and animal protein. In the 1980–2010 period, the rate of increase of food production in the bloc was double that of the rate experienced at a global level (Reca, 2012). This dominant position has been maintained in recent years, and has even strengthened in relation to some products. With nearly 139 million metric tons produced in 2012/2013, the MERCOSUR bloc is now the main producer of soybean worldwide, accounting for 52% of total global production (United States Department of Agriculture, 2014).

Over the past four decades the South American region recorded the fastest growth in soybean of any region in the world. The cultivated area increased by a magnitude of 20 times between 1991 and 2010 (see Table 1). The process has involved a fundamental reorganization of territory in South America from the point of view of production. The expansion of soybean has involved the expansion of the total area of land under agricultural production, the replacement or displacement of other agricultural products, and the cultivation of land formerly under natural forests. Expansion has also had major impacts on the agrarian structure of many countries, since soybean is produced mainly by large-scale and medium-scale farmers: although about 73,000 farmers in Argentina planted this crop in 2008 (around a quarter of the total number of producers), only 6%, with farms over 500 hectares, produced nearly 50% of the total yield (ONCAA 2008, cited in Regunaga, 2009). In Brazil, the concentration of soybean production in large landholdings is even higher. Only 4% of all farms (217,000) produced soybean in 2006, yet those with farms over 500 hectares representing 5% of total farm units, accounted for 2/3 of soybean production (Wesz, 2014a).

Changes in the world soybean market are important for explaining the expansion of soybean production in the MERCOSUR countries. In the 1960s and 70s there was a temporary moratorium on US soybean exports[6], together with a global shortage of Peruvian anchovies, the main source of fishmeal used by the animal feed industry at that time. From 2000 onwards, the increased demand from

Table 1. Changes in the total area of arable land and in the area planted with soybean in MERCOSUR countries, 1991–2009.

Country	Year	Arable land (in million hectares)	Soybean-planted area (in million hectares)	Per cent of arable land with soybean
Argentina	1991	26.4	5	18.94
	1995	27	6	22.22
	2000	27.9	10.66	38.21
	2005	29.5	15.39	52.17
	2009	31	18.34	59.16
Bolivia	1991	2.11	0.19	9.00
	1995	2.5	0.43	17.20
	2000	3	0.62	20.67
	2005	3.81	0.93	24.41
	2009	3.74	0.9	24.06
Brazil	1991	52	9.62	18.50
	1995	58.06	11.68	20.12
	2000	57.7	13.64	23.64
	2005	61	22.95	37.62
	2009	61.2	21.75	35.54
Paraguay	1991	2.15	0.55	25.58
	1995	2.6	0.74	28.46
	2000	3.02	1.2	39.74
	2005	3.46	2	57.80
	2009	3.8	2.52	66.32
Uruguay	1991	1.26	0.02	1.59
	1995	1.29	0.01	0.78
	2000	1.37	0.01	0.73
	2005	1.3	0.28	21.54
	2009	1.88	0.58	30.85
MERCOSUR	1991	83.92	15.38	18.33
	1995	91.45	18.85	20.61
	2000	92.99	26.13	28.10
	2005	99.07	41.54	41.93
	2009	101.62	44.09	43.39

Source: Catacora Vargas et al. (2012)

emerging countries for animal feeds and the expansion of biofuels were the key determinants of increased production of soybean (Costantino & Cantamutto, 2010).

While Brazil and Argentina together hold 91% of MERCOSUR's soybean acreage, there are significant differences in the character of the soybean value chain in each country, and differential degrees of specialization that affect their respective vulnerability to external shocks. In Brazil, the 1996 Kandir Law eliminated a tax on exports of raw materials and kept the tax burden on industrialized products (Wesz, 2014b). Consequently, in 2010 65% of soybeans were exported, 65% of them non-processed (ABIOVE, 2015). While Argentina has re-established export taxes on grains since 2002, the tax regime benefits value-added products and encourages local processing. In 2010 77% of soybeans were exported, 57% as flour and 12% as oil (CIARA, 2010). National companies have a greater weight in Argentina than in Brazil, controlling over 30% of exports of flour and soybean oil (Wesz, 2014b).

In Argentina the soybean complex provided 30% of all exports in 2015, while it accounted for only 15% of exports in Brazil. However, the contribution of soybean to total exports has been increasing in both countries (see Figure 1). The sector strengthened its importance in Argentina as it supplied foreign currency in a context where access to international credit was restricted after the country's debt default of 2002 (Craviotti, 2015). Fees from major agricultural exports are also important for the state budget.[7]

From the geographical point of view, soybean cultivation began in Argentina in the most fertile areas of the Pampean region. Although production of the crop has expanded to the northwest and northeast of the country, the Pampas still represent 85% of the total soybean acreage (SIIA, 2011). In Brazil soybean developed first in the south (in the states of Rio Grande do Sul and Paraná), and expanded later in the centre-west and centre-north of the country, in the *Cerrados* zone and parts of the Amazon region. These areas thus displaced production in the 'traditional' southern regions, which account for only 37% of the cultivated area today (APROSOJA, 2011). A few decades ago, the *Cerrados* were not suitable for grain production, but the involvement of different government agencies was critical for the opening up of this zone for agribusiness investment (Gras, 2013). Thus, the state-owned company EMBRAPA (*Empresa Brasileira de Pesquisa Agropecuaria*)

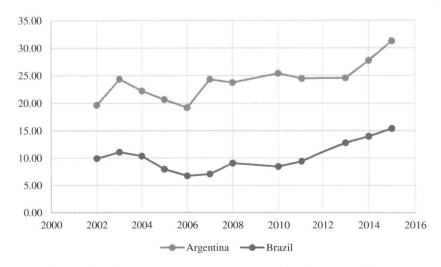

Figure 1. The contribution of soybean to exports in Argentina and Brazil (as a % of total exports). Source: The author, based on data from INDEC (2002–2015) and ABIOVE.

developed fertilizers that allowed soil acidity in the savannah to be reduced, and also soybean varieties adapted to this region. Government also invested in roads and irrigation infrastructure, as well as offering low-interest loans to producers. Some of the states of the *Cerrados* also have tax incentive programmes designed to attract agricultural producers and business ventures (Campos Mesquita & Lemos Alves, 2013).

In the case of Argentina, public support of this magnitude for the development of soybean did not exist, but the emergence of a technological package, based on the combination of transgenic varieties resistant to glyphosate and no-tillage, was crucial. A set of public policies enabled Argentine producers to access this key package. On the one hand, the official approval of glyphosate-resistant soybeans occurred almost simultaneously with their release in the United States (1996). The area sown with genetically modified (GM) soybeans increased from less than 1% of the total soybean planted area in 1997 to more than 90% in 2002, a rate of adoption then considered higher than in the United States, the first country to introduce this technology (Trigo & Cap, 2003). Today, GM soybeans cover 20.3 million hectares in Argentina (ARGENBIO, 2016), nearly 100% of an estimated area of 20.6 million hectares under the crop.

The widespread adoption of GM soybean was also facilitated by the self-pollinated nature of the seed, whose reproduction does not alter its initial characteristics, and the strategy followed by the seed companies for rapid dissemination of this technology. The first application to the National Advisor Comission for Agricultural Biotechnology (Comisión Nacional Asesora de Biotecnología Agropecuaria, CONABIA), a public body composed of representatives of the state, industry and the scientific community, to conduct field trials of GM soybean varieties was presented by the company Nidera in 1991. Later, Monsanto could not patent the gene that confers tolerance to glyphosate because the technology had already been 'freed'.[8] This, together with the possibility of using farm-saved seeds (a right recognized by the Argentine legislation in coherence with its participation in UPOV 1978) allowed the rapid diffusion of the new technology. Besides, the lower operating costs that the new package initially enabled was a key motivation for its adoption, given that economic policies implemented in a context of declining international grain prices pushed local farmers towards cost-saving strategies (Craviotti, 2002).

The widespread adoption of GM soybean contributed to the massive expansion of the area under the crop. In contrast, Brazil did not allow GM seeds until 2003, a factor that explains its later expansion when compared to Argentina. Nevertheless, producers from the southern states of Brazil obtained seeds illegally from Argentina. Before the final approval of the glyphosate-resistant soybean by Law 11,105 in 2005 (the Biosafety Law), the government had already authorized grain marketing of GM soybeans in 2003/2004 and 2004/2005 through a series of provisional measures (Brieva, 2006). Ten years later, it is estimated that GM soy occupies 93% of the area in Brazil planted to this crop (Celeres, 2014).

In Argentina, after an impasse in the 1999–2003 period, the administrations of Nestor and Cristina Kirchner resumed the path of support for biotechnology with the approval of new traits and the enactment of Law 26,270 for the Promotion of Development and Production of Modern Biotechnology in 2007 (Idigoras, 2013). Biotech approvals between Argentina and Brazil are now being harmonized; for instance, the two countries have entered into joint negotiations with the Chinese government for the approval of new soybean seeds that embody 'stacked' GM traits (insect resistance and glyphosate tolerance). This strengthens a trajectory of agrarian development, in which flex crops represent an important basis for re-negotiating the role of both countries in the international division of labour.

Argentinian *multilatin* firms in the soybean sector

In addition to public policy, another important factor underlying the massive expansion of GM soybean in the MERCOSUR region has been the emergence of *multilatin* companies whose accumulation strategies are strongly founded on this profitable crop. Their growth is obviously linked to globalization, which involves financial and trade flows across national boundaries, as well as the strengthening of global value chains (Schorr & Wainer, 2014). Although today Argentina diverges from other Latin American economies in relation to the magnitude of outward flows of investment – they range from 700 to 2000 million dollars in the 2007–2015 period, while in Chile and Mexico they have peaked 22,000 million in some years – the country is undoubtedly part of this trend (Figure 2).

Different phases can be identified in the internationalization of Argentinian firms. In the early twentieth century, and in line with the country's role at the time as a leading producer of wheat, corn and beef worldwide, a limited number of local companies were the first firms not based in developed countries that established overseas facilities. This trend continued during the import substitution stage of the 1940–1970 period, the ventures concerned being of relative low levels of investment, but gained new momentum in the 1990s, a period marked by the implementation of neoliberal policies in Latin America. Increases of scale of production and degree of internationalization were required in a context of increasing competition (Kosakoff & Ramos, 2010). Also, the regional integration process that began in 1986 between Brazil and Argentina, and which extended to other countries in South America through the creation of MERCOSUR in 1991, removed barriers to trade and facilitated this process (Belik & Rocha Dos Santos, 2002). Some Argentinian firms expanded their operations to other countries in this stage. It should be noted, however, that leading local firms were also bought by foreign capital, increasing levels of foreign ownership (Kosakoff & Ramos, 2010; Schorr & Wainer, 2014).

With the marked devaluation of the Argentinian currency in 2002, a new phase of internationalization began. Some Argentinian firms started operations in other MERCOSUR countries focused on the soybean value chain, propelled by the growth of global demand for the product, especially from China.

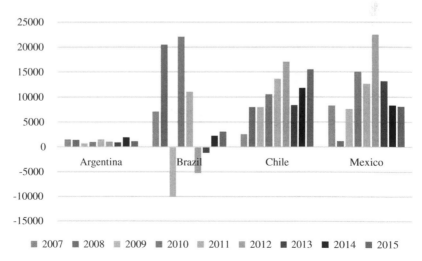

Figure 2. Outward flows of foreign direct investment in selected Latin American countries. Source: The author, based on data from United Nations Conference on Trade and Development (2005, 2010, 2016).

Over the long run, however, the trend is declining levels of outward investment by Argentinian firms in general, when compared with those from other major economies in Latin America (ECLAC, 2006; Finchelstein, 2012). This fact highlights the diverging trajectory of those Argentine firms that have internationalized their operations in recent years through investments in different parts of the soybean chain.

'Sowing pools' and seed firms stand out in the map of actors that have pursued this strategy. The firms that form sowing pools do not necessarily own the land that they operate but, rather, lease it on a short-term basis under various contractual arrangements while outsourcing farm tasks through contracting. The biggest pools also have fluid connections with financial markets. Sowing pools have developed mainly in grain agriculture, and have soybean as a main component of their strategy. Regional expansion allows them to spread climatic and political risks, to diminish operating costs, and to gain access to key resources such as cheap land (Bisang & Anllo, 2014; Gras & Sosa Varrotti, 2013; Manciana, Trucco, & Piñeiro, 2009). Some leading sowing pools have also established partnerships with foreign capital to facilitate large-scale investment (Murmis & Murmis, 2012).

In the case of seed firms, there are technical aspects that favour the move to international operations: germplasm can be transferred between different areas and countries, and testing in different natural environments, rather than a narrow range of ecosystems, enables identification of the most suitable materials for development (Jacobs & Gutierrez, 1985).

At a global level, the internationalization of the seed industry began in the 1970s and led to a global restructuring. The number of mergers and acquisitions increased greatly in the 1990s, when hybrid seed companies were integrated into the agro-chemicals industry. Concentration in the global seed market saw the four leading firms increasing their share from 21% in 1994 to 54% in 2009 (Fuglie et al., 2011).

Soybean varieties are self-pollinating, however, and this opened space for the development of local actors in the seed industry. In Argentina, systematic research on this crop was initiated about five decades ago, with state agencies and national firms testing the adaptation of materials originating mainly from the United States (Jacobs & Gutierrez, 1985). As years passed, plant breeding by local operations began to take off, public sector organizations (such as the National Institute for Agricultural Technology, INTA) lost ground, as local companies began to develop their capacity to undertake germplasm crossings. At this stage transnational corporations (TNCs) did not display much interest in soybean (Brieva, 2006). Since they are self-pollinating, harvested soybean seed can be used by farmers without any change in their qualities. This condition not only extends the farmers' room for manoeuvre but also limits the ability of the private sector to increase the price of seeds. Thus, foreign companies focused on hybrid seed (for crops such as maize, sunflower and sorghum) where there is a 'natural' barrier against multiplication and farmers are forced to buy seeds every planting season.[9]

Today, the Argentinian soybean seed market is supplied almost entirely with genetic material of local origin (Figure 3), and national firms – such as *Don Mario, Santa Rosa, and Agseed* – are important suppliers within this market (Figure 4). However, in a context dominated by GM soybean, these local breeders establish agreements with biotech firms (which are mostly TNCs) to add transgenic traits to their varieties. As a result, they remain rather invisible in studies of the seed industry, since their varieties tend to be 'subsumed' within the accounts of large biotech companies.

The evolution of the Argentine register of cultivars over the last 15 years shows a marked mobility of companies in the soybean sector, although some of them hold a more stable position. Among the

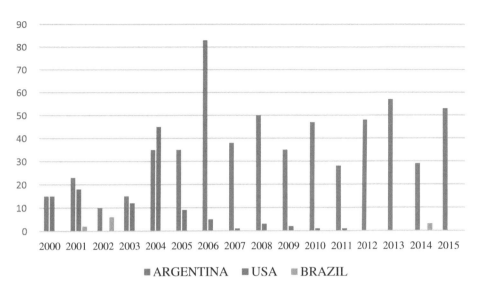

Figure 3. Argentina. Origin of soybean varieties. Source: The author, based in the National Register of the Property of Cultivars.

latter, *Don Mario* holds a leading role together with *Nidera*, a company of Dutch origin that has been recently bought by the Chinese state-owned firm COFCO. These two firms control 90% of the market, and produce the entire range of the varieties grown in the country.

A handful of these seed companies – *Nidera, Don Mario, Santa Rosa, Agseed* (the later three of local origin)- have also started to deploy an internationalization strategy in the last decade, with soybean varieties as flagship products, and Brazil as a key destination. According to a key informant from the public sector (author interview, 2014), their early expansion was related to the fact that:

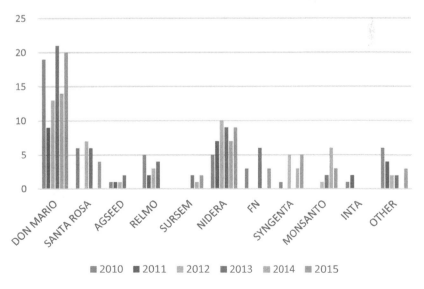

Figure 4. Argentina. Registered soybean varieties by firms and institutions. Source: The author, based in the National Register of the Property of Cultivars.

Argentina pioneered genetic development [...] the other issue is that Brazil at that time did not admit transgenic seeds, and the transgenic materials that these companies could provide had much know-how and things to offer with a different germplasm base.

These firms view Brazil as the country with the highest potential for investment, since it is the second largest producer of soybean after the United States, and contains vast areas in the Cerrados biome where production can be significantly expanded.[10] In addition, according to key informants, farmers of this region do not engage in the practice of saving their seeds.

Another factor that attracts the interest of seed companies in Brazil is its stronger recognition of intellectual property rights, when compared with Argentina. Although the seed laws in the two countries are similar in some respects, they differ in others – notably in relation to those that allow seed firms to capture value (Table 2). Both countries do not allow patents for plants, and they recognize the rights of rural producers to save seeds, and of breeders to use existing protected varieties to develop new ones. However, in the case of Brazil, if a new variety is distinguishable but derived predominantly from an original, protected variety, the authorization of the owner of the original cultivar is required for purposes of marketing. The right of farmers to save seeds is also limited to the second generation of seeds purchased in the formal market.

In Brazil, on the other hand, seed firms have succeeded in co-opting some producers' organizations and have secured private contracts to collect royalties (Filomeno, 2013; Scoones, 2008), a practice that has been resisted in Argentina. All these factors facilitate successful accumulation strategies by seed firms in Brazil.[11] The purchase of certified seed is higher, bringing in an estimated annual income of US$1.2 billion, in contrast with US$240 million in Argentina (*La Nación*, 4 September 2014). Government agribusiness policies seem to be an important element at play for Brazil, as one key informant said, 'all the players you meet speak of the importance of agriculture in Brazil, and defend it' (author's interview, 2014). The so-called 'business climate' also matters – local interest rates are lower in Brazil than Argentina, affecting the cost of external financing for firms expanding their investments.

Taking into account this general background, the evolution and mode of operation of a leading Argentinian seed firm will be analysed in the next section. Some comparative material on other seed firms is also discussed. With only a small number of firms dominating the market, the strategies of these firms, within the prevailing regulatory framework, play a key role in shaping the dynamics of the soybean seed market (Fuck & Bonacelli, 2007).

Table 2. Prevailing legal framework in Argentina and Brazil.

		Plant variety protection				
	Year of adhesion to UPOV	Farmers' privilege	Breeders' exemption	Essentially derived variety[a]	Protection period	Possibility of patenting plants
Argentina	1994 (1978 Act)	Yes	Yes	No	20 years	No
Brazil	1997 (1978 Act and elements of 1991 Act)	Yes[b]	Yes	Yes	15/18	No

Source: The author, based on Salles-Filho et al. (2011) and Wilkinson and Castelli (2000).

[a]In varieties that retain most of the genetic components of another variety, the original breeder's authorization is required for marketing of the variety.

[b]Small, family farmers are allowed to multiply seeds for donations or in-kind exchange with other small farmers. The farmers eligible are those who farm a plot of land as an owner, squatter, renter or sharecropper; employ no more than two hired persons on a permanent basis; do not hold an area greater than four 'fiscal modules', as set out by existing legislation; earn at least 80% of their gross annual income from farming, cattle-raising or extractive activities; and live on the farm or in a nearby urban or rural settlement.

An internationalized seed company of Argentinean origin

As argued above, the soybean seed sector in Argentina is heterogeneous, with a spectrum of producers that includes global companies, firms of national capital, and producers' cooperatives that began their activities in a rather modest way. One of the two leading firms in the country is owned by local partners, and was created in 1982 by a group of friends who together invested a small sum of money (US$15,000) in grain production on rented land. In the beginning, they rented plots in an in-kind basis, an arrangement that allows the farmer and the landowner to share risks.[12] Meanwhile, they performed testing of soybean varieties for a local, non-profit technical organization. This allowed them to see the advantages of the short-cycle varieties in which the company would later specialize.[13]

Three years after the firm initiated the production of seeds, using an American short group variety, then of public origin. Taking these varieties as its starting point, it developed a niche previously unoccupied by other seed companies. Short cycle varieties expanded in the Pampean region because of their advantages: they bloom before longer cycle varieties, they are resistant to the development of certain diseases (Sclerotinia), and their yields are higher. However, as the critical period of crop development falls earlier than other varieties, the risk of crop failure is increased in drought periods. However, no-tillage techniques also began to be developed around this time in Argentina, and these, by helping to preserve soil moisture, have mitigated the impact of drought.

After 1989 the company was able to import machinery and various seed varieties into Argentina. This followed the coming to power of the neoliberal government of President Carlos Menem, with its policy of strongly encouraging foreign investment, coupled with the establishment of 'one to one' parityup between the Argentine peso and the US dollar. The firm continued to test imported seed varieties and strengthened its presence in the local market, thanks to its policy of forming alliances with TNCs. Thus, a global grain trader bought seeds from the company and sold them to farmers, also offering financial assistance to the farmers interested in buying these seeds. This partnership lasted from 1992 to 1995, and was replaced in 1996 by an agreement with Monsanto, which allowed the company to market its seeds through Monsanto's distribution networks, and to employ the GM seeds (resistant to glyphosate) to develop its own varieties through germplasm crossings. These seeds started to be known as 'RR' because of the commercial name (Round-up Ready) of the herbicide.

The agreement with Monsanto also brought indirect competitive advantages to the seed firm. In 1998 it bought another local company whose value in the market had diminished because it could not establish an arrangement with Monsanto regarding access to the transgenic trait. This acquisition allowed the firm to expand its portfolio of varieties (until then limited to the Pampean region), and to achieve national coverage, following expansion of the area planted to soybean. At this stage, the firm changed its status from that of a limited liability company to a corporation (Don Mario, 2005).

The devaluation of Argentina's currency in 2002 radically altered the context and greatly increased the cost of royalties to be paid to owners of patents. In addition, Monsanto adopted a much tougher stance in relation to the Argentinian regulatory framework for seed production, resulting in increased pressure on other seed companies. In 2004 Monsanto announced its withdrawal from the soybean industry in Argentina, in retaliation for the country's denial of its application for a patent on the glyphosate-resistance technology, and filed lawsuits in countries importing soybean products from Argentina (Filomeno, 2013). The national firm decided to strengthen its in-house breeding programme and to create its own sales network, since its agreement with Monsanto had ended. In fact, it can be argued that the withdrawal of the biotech company from Argentina, together with the boom of soybean production, in fact promoted the national firm's growth.

At this stage, the company affirmed its vision of becoming an industry leader in soybean genetics. In the 1996–2000 period, it had registered only four soybean varieties in the country, but over the next five years the figure rose to 24 (*Infobae*, 27 October 2006). The firm has registered about 50 varieties since 2000. However, upgrading of its capacity to develop specific traits is not part of its plans for the near future. It is estimated that in the case of soybean, the development of a transgenic trait requires about 16 years and a 136-million-dollar investment (Rocha & Villalobos, 2013).

In the period since 2000 the company has not only increased its turnover but also bought up other firms in the seed sector. It has developed a holding structure which at present includes 10 related firms in Argentina and a similar number in neighbouring countries, as well as in the United States, and South Africa. It has created companies specifically to make finance available to its suppliers, and placed securities on financial markets in order to raise capital at low interest rates. The possibility of registering the company on the stock exchange is under consideration by firm partners. These developments can be explained by the increasingly large amount of capital required for producing and processing seeds in Argentina, as well as for the investments required to expand into Brazil.

The *modus operandi* of the firm illustrates its character as a *network within networks* (Dicken & Malmberg, 2001). It has developed an external network of alliances with firms which are leaders in the provision of agri-food inputs worldwide. It has established agreements with biotech companies for the use of transgenic seed traits, and with other companies for co-investment in new enterprises, such as a joint venture with Louis Dreyfus for the production and sale of corn seed, and another with Dow for operation of a hybrid corn processing plant, both in Brazil. The Argentine company and Dow also have a partnership in Bolivia, for the production and marketing of soybean varieties.

The firm's initial alliance with Monsanto has given way over time to a more diversified network of partners, which includes other TNCs. Currently, the attractiveness of partners seems influenced by two considerations: firstly, the firm's need to strengthen its position in Brazil; secondly – and as explained below – tensions regarding the capture of the benefits associated with the 'new' generation of transgenic seeds.

Just as the emergence of glyphosate tolerant soybeans 10 years ago radically altered the map of key actors in the seed industry, similar processes have recently arisen with the advent of 'stacked' GM traits in soybean seeds. In this case, Monsanto has aimed to ensure the collection of royalties from the very beginning, and has entered into agreements with local breeders, grain processors and exporters to test for the presence of these traits in soybeans delivered by farmers. The biotech company has also tried to have the value of farmers' royalties vary in accordance with the productivity increases obtained.

All of these strategies of Monsanto's were questioned by farmers' organizations in Argentina, resulting in lower than anticipated sales of soybean seeds with stacked GM traits. In turn, this has caused problems for the Argentinian firm, which had bet heavily on these varieties. Also, and from the breeders' perspective, the technology fees established by Monsanto tend to overvalue genetically engineered traits at the expense of the contribution of germplasm to the increase of productivity, and this has generated tensions between partners. These arise from the asymmetries in access to resources and position in the global seed industry that exist between breeders of national origin and biotech TNCs.

So, in the last years the Argentinian company seeks to maintain a degree of technological independence through strengthening its in-house research programme for breeding non-GM soybeans. Other local firms engage in similar strategies. This increases their flexibility, as it enables the companies to add to its own seeds transgenic events developed by other biotech firms, and enables to develop new niche markets for non-transgenic soybean varieties. As stated by a firm representative:

> We have conventional varieties that were in existence before the [GM]traits [...] it is a very strong pro-gramme [of breeding] because if a new biotechnology comes up, we insert it there. Because many times biotech companies do not have agreements, they don't let you put the trait by Dow on the RR of Monsanto. [The conventional program] now starts to be more valuable because there are resistant weeds; RR no longer has the same value. (Author's interview, 2014)

It should be added that the leading Argentinian firm has also created a subsidiary for producing non-GM soybean for export, either directly itself or through contracts with outgrowers. Although the overall volume managed is relatively small, it has been growing over time, so this firm has become the main exporter of non-GM soybean from Argentina (*Infocampo*, 17–23 October 2014).

However, the subordinate position of local breeders in relation to biotech corporations is not necessarily replicated in their relationships with local actors. The leading Argentine firm outsources an important part of its seed production to diminish operating and financial costs, and to disperse risks amongst many actors. Outsourcing is achieved through a variety of mechanisms, including agreements with *co-operators,* who buy basic seeds, and then multiply, classify and sell them, paying royalties to the company; and agreements with *multipliers* (farmers), to whom the firm sells basic seeds and then buys from them multiplied seeds at harvest. For activities related to the 'industrial' phase of seed production, the company has its own seed classification plants. The company also hires out processing services to a range of third parties.

Another key strategy employed by Argentinian soybean seed firms over the last decade has been to expand abroad. This began with the commodities 'boom' of the 2000s, and was initially deployed in Argentina's neighbouring countries. It also encompassed South Africa – by far the dominant GM seed producer in Africa (Wield, Chataway, & Bolo, 2010) – considered as a possible 'springboard' to other countries of the continent. The United States is one of the most recent destinations, while other countries such as Ukraine, Russia and China are also being considered (*Perfil*, 3 February 2014). However, Brazil is viewed by these *multilatin* seed firms as the market with the greatest potential, as outlined above. On the other hand, the progress made by seed firms of Argentinian origin in the development of varieties and especially of short-cycle soybeans was probably a competitive advantage, as indicated by a key informant of the public sector. The manager of a seed company in southern Brazil confirms this perspective, stating that *Argentine cultivars had already advanced in research with RR, and occupied space quickly. Brazil used materials with a later maturity cycle], with many leaves and a certain growth habit. These materials ended up suffering diseases, while the Argentine cultivars, with less leaves, more efficient and with a shorter cycle [of maturity] had a great advantage over the Brazilian cultivars (La Nación, 9 September 2014).* These indeterminate short varieties are particularly suitable for an annual double-cropping scheme.

To begin operations in these new territories, seed firms usually establish agreements or partner-ships with local entrepreneurs who facilitate access to key resources (knowledge of locations suitable for testing varieties, links to seed growers, etc.). In Brazil, the leading Argentinian firm chose to develop its activities through a partnership with local seed firms, which was created in the same year that the Brazilian government authorized the commercialization of GM soybeans in the state of Rio Grande do Sul (2003). As a company representative explained:

> They helped us to understand the Brazilian farmer, the business culture of Brazil. In this business, seed varieties must be registered, so there is a link with the state, and Brazil has a complex bureaucracy. (ACDE, 2012, p. 13)

Specifically, this partnership allowed the firm to register many varieties in just a few years; it has already registered about 60 in Brazil. Yet disagreements on the research focus precipitated changes in the company, increasing the share of the Argentinian partners. In 2008, they bought up all of the firm's stock, and one year later they created a completely new company in Brazil.

Despite its short history in Brazil, over half of the firm's staff and its largest laboratories are located there.[14] In 2013 the Brazilian market accounted for 51% of the group's net income (Financial Statements, 2014). The firm estimated that it held 24% of the Brazilian soybean market, with a 55% share of the market in the southern states of Brazil (Financial Statements, 2014).

As is the case with other Argentinian seed companies that have expanded their operations abroad, the leading firm also undertakes local research. In 2014, it reported that it had undertaken research trials in 160 localities across all America, Africa, Asia and Europe. Argentina itself was the location for the largest number of trials, followed by Brazil. The research department of the firm works as a single multinational team, gathering information from all locations (author's interview, 2014).

Along with internationalization, the company developed the concept of 'Yields with no Borders'. This is based on the notion that political divisions are of secondary importance to the company, only latitude and climate matter, and accordingly, the focus is on the most suitable varieties for each territory.

> When looking at a South American map, [the members of the company] see the South American region globally. Although there are different cultures, soybean does not face political boundaries, soybean is only one. (Don Mario, 2013, p. 98)

The similarity of this vision to the one introduced in 2005 by Syngenta, which refers to MERCOSUR as the 'United Republic of Soybean', is clear.

Conclusion

This article has focused on the development of *multilatin* firms in the soybean seed sector of countries in the MERCOSUR bloc, with a focus on Argentinian companies. Viewed from a long-term perspective, a key feature of the Argentinian firms that have expanded into other countries has been their interest on the production of commodities, or in activities directly linked to such production. Yet their present involvement in key stages of the South American soybean chain as a whole is a relatively new phenomenon, which cannot be dissociated from the role played by the country in the global trade, which has a strong focus on 'flex crops', particularly transgenic soybean.

Seed firms do not necessarily engage in land grabbing and control. On the contrary, their strategies demonstrate the importance of intangible assets such as brands and intellectual property rights (Pritchard, 2000) in attempts to amplify their influence on other actors and thus strengthen processes of accumulation. These features pose challenges to contemporary studies of agrarian change, which tend to emphasize control of tangible assets.

The strategies adopted by these firms also show the existence of South-South flows of capital, that may indicate the incipient formation of transnational elites grounded in global circuits of accumulation (Robinson, 2015). These developments are not yet stabilized: the procurement of Nidera by COFCO and of Syngenta by ChinaChem imply the emergence of new flows of capital from BRICS countries, that aim to strengthen their position within key components of agri-food chains.

Until now *multilatin* firms have adopted a regional rather than global strategy, and have taken advantage of processes of market integration facilitated by trade agreements. In the case of Argentinian seed firms, they have also considered countries that do not belong to MERCOSUR as possible sites of expansion. In terms of the paradigm posited by Dunning (2000), the foreign activities of these firms have been driven by a specific set of advantages: their dominant position in the breeding of transgenic soybean varieties, facilitated by early approval of the glyphosate-resistant event, and underpinned by the specific conditions enabling its diffusion in Argentina. However, locational advantages, and particularly the differences between countries in relation to regulatory frameworks, have also been key, enabling value capture and accumulation. Other aspects, such as the growth potential of the 'new' soybean areas (such as the Brazilian Cerrado) have also been important factors.

In the case of the leading Argentinian seed firm, organizational innovations have played a key role in the firm's development and its internationalization. These are evident in its ability to build an effective internal network for the production and marketing of seeds which, through outsourcing, also enables flexibility and risk avoidance. Connections to global corporations, on the other hand, have enabled access to key resources and investment opportunities, while relationships with selected local actors have facilitated the firm's growing role in new settings.

From another point of view, this multifaceted network that involves public and private actors in the soybean chain, as well as technical objects (GM seeds), enables an alignment of interests and secures a specific technological path (Latour, 1994), that constrains its reversibility. However, as illustrated above, there are currently tensions between different fractions of capital of different origin, control of access to resources and position within the network, a situation that could ultimately bring about unexpected results. The move by national breeders to resume crossings of non-transgenic soybean varieties, although motivated primarily by the opportunities opened up by the multiple licence agreements that govern biotechnology, could indirectly help to broaden the scope of possible choices. Yet, the latter seem to be confined to soybean, ruling out a general move towards diversification.

Notes

1. Flex crops are agricultural products that can be used as food, feed and biofuels, and which can be changed flexibly according to circumstances (Borras, Franco, Kay, & Spoor, 2011). In the MERCOSUR countries maize, sugarcane and soybean are good examples.
2. Following Bastian and Soihet (2012) *reprimarization* can be defined as an increase of the share in exports of primary and manufactured products with low value added and/or low technological content. *Neo-extractivism* is a model of development based on the appropriation of nature, which sustains a barely diversified productive structure and involves the insertion of a country into the world economy mainly as a provider of raw materials.
3. A *multilatin* or *global latina* firm has been defined as a company with its origin in a Latin-American country that has value-added operations outside its country of origin (Cuervo Cazurra, 2010). In this article, I follow the more restrictive definition suggested by ECLAC (2006, p. 63), which considers 'trans-latins' as emerging Latin American transnational firms that have made direct investments outside their home countries.
4. MERCOSUR (Common Market of the South) was created in 1991 when Argentina, Brazil, Paraguay, and Uruguay signed the Treaty o Asunción, establishing the free movement of goods, services, and factors of production between countries.
5. The term 'internationalization' of firms adopted throughout the article refers to the development of international operations, basically investment in foreign countries.
6. Soya oil was originally developed with US state subsidies to supply the margarine industry. After World War II it became more important as a joint product of soya meal for the intensive livestock industry. With the Soviet purchases of the early 1970s, prices soared and the US government feared domestic shortages, placing a temporary embargo on soybean exports. Brazil and Argentina cut into world exports (Friedmann, 1992).
7. The income from agricultural export taxes (mainly from soybeans) has oscillated between 13% and 6% of total state income in the 2008–2014 period (http://www.mecon.gov.ar/sip/).

8. Nidera bought Asgrow Argentina in 1988 and began its activity as Nidera Seeds. Asgrow International had access to Monsanto's technology through an agreement between the two companies secured in the United States in the late 1980s (Brieva, 2006).
9. Despite these factors, over the past decade TNCs have developed a greater interest in the Argentine soybean seed market, a process that is linked to the possibility of changes in the regulatory context such as a shift to a much stricter intellectual property regime.
10. The Cerrado biome is located in the central part of the country and covers approximately 204 million hectares (or 24% of Brazil's entire land area). An estimated 40–50% of the Cerrado is under productive use, and by 2008 accounted for 59% of Brazil's coffee production, 55% of its beef, 54% of its soybean, 28% of its corn, and 18% of its rice (Trigo, Cap, Malach, & Villareal, 2009).
11. Seed companies favour changes in the Argentinian legal framework to ensure the recovery of intellectual property rights, thus restricting the right to save seeds to certain categories of farmers. In 2003 a state initiative intended to adopt a new law to govern the production and sale of seeds did not succeed, but a new attempt in this direction began in 2012, after the approval of a second generation of transgenic seeds with stacked traits. Some seed companies have also entered into private contracts with producers that allow them to collect extended royalties for farm-saved seeds. However, the system has limited coverage due to the resistance of Argentinian producers' organizations (Filomeno, 2013). According to a seed firm representative, 38% of the seeds sown in Argentina have recognized intellectual property rights to them (considering the sale of both certified seeds and royalty payments). In Brazil, this reaches 60%; in Uruguay, 100%; in Bolivia, 65%; and in Paraguay, 40%.
12. This type of arrangement implies that the tenant gives the landowner a percentage of the crop harvested instead of a cash payment in advance.
13. Soybean varieties are classified for their morphological growth habit, and for their day length and temperature requirement to initiate floral or reproductive development. A short-cycle variety matures in 90 to 100 days, which is two to three weeks earlier than the traditional varieties, allowing farmers to plant a second crop after the soybeans are harvested.
14. These facilities are already undertaking 3,000,000 analyses per year, while in Argentina they can only process about 300,000 samples annually.

Acknowledgements

The author would like to thank technicians, government officials, and firm representatives for their participation in personal interviews. The paper greatly benefited from the comments made by anonymous reviewers and the editors of the Special Issue. The author would also like to thank BRICS Initiatives in Critical Agrarian Studies (BICAS) for providing funding to carry out this research.

Disclosure statement

No potential conflict of interest was reported by the author.

References

ABIOVE (Brazilian Association of Vegetable Oil Industries). (2015). *Brasil. Exportaçoes do complexo soja*. Retrieved from http://www.abiove.org.br

ACDE (Cristian Association of Business Leaders). (2012). *XV Encuentro Anual de ACDE*. June 26. Buenos Aires.

APROSOJA (Brazilian Association of Producers of Soybean). (2011). *Serie Historica de area plantada. Safras 1976/77 a 2014/15*. Retrieved from http://www.aprosoja.com.br

ARGENBIO (Argentine Council for Information and Development of Biotechnology). (2016). *Argentina: Cultivos GM campaña 2015/2016*. Retrieved from https://www.argenbio.org/adc/uploads/imagenes_doc/planta_stransgenicas/2016/Argentina_cultivos_GM_campana_2015_2016.pdf

Argentinian Seed Association. (2013). *Importancia del sector semillero en la economía argentina*. Buenos Aires: Author.

Bastian, E., & Soihet, E. (2012). Argentina y Brasil: Desafíos macroeconómicos. *Problemas del Desarrollo, 171*, 83–109.

Belik, W., & Rocha Dos Santos, R. (2002). Regional market strategies of supermarkets and food processors in extended Mercosur. *Development Policy Review, 20*(4), 515–528.

Bisang, R., & Anllo, G. (2014). *Impactos territoriales del nuevo paradigma tecno-productivo en la producción agrícola argentina. Documento de trabajo 5*. Buenos Aires: Instituto de Economía Política de Buenos Aires.

Bonny, S. (2014). Taking stock of genetically modified seed sector worldwide: Market, stakeholders, and prices. *Food Security, 6*(4), 525–540.

Borras, S., Franco, J., Kay, S., & Spoor, M. (2011). *El acaparamiento de tierras en América Latina visto desde una perspectiva internacional más amplia*. Santiago: FAO.

Brieva, S. (2006). *Dinámica socio-técnica de la producción agrícola en países periféricos: configuración y reconfiguración tecnológica en la producción de semillas de trigo y soja en Argentina desde 1970 a la actualidad* (PhD thesis). FLACSO, Buenos Aires.

Campos Mesquita, F., & Lemos Alves, V. (2013). Globalización y transformación del paisaje agrícola en América Latina: las nuevas regiones de expansión de la soja en Brasil y la Argentina. *Revista Universitaria de Geografía, 22*(2), 11–42.

Catacora Vargas, G., Galeano, P., Zanona Agapito, S., Aranda, D., Palau, T., & Onofre Nodari, R. (2012). *Soybean production in the Southern Cone of the Americas. Update on land and pesticide use*. Cochabamba: Vimegraf.

CELERES. (2014). *Informativo biotecnologia*. Retrieved from http://www.celeres.com.br

CIARA (Argentinian Oil Industry Chamber). (2010). *Origen y aplicación de los cinco principales granos oleaginosos de Argentina*. Retrieved from http://www.ciara.com.ar

Costantino, A., & Cantamutto, F. (2010). El Mercosur agrario: ¿integración para quién? *Iconos, 38*, 67–80.

Craviotti, C. (2002). Pampas family farms and technological change: Strategies and perspectives towards genetically modified crops and no-tillage systems. *International Journal of the Sociology of Agriculture and Food, 10*(1), 23–30.

Craviotti, C. (2015). Argentina's agri-food transformations in the context of globalization: Changing ways of farming. In A Bonanno & L. Busch (Eds.), *Handbook of the international political economy of agriculture and food* (pp. 78–96). Chetelham-Northampton: Edward Elgar.

Cuervo Cazurra, A. (2010). Multilatinas. *Universia Business Review, 25*, 14–33.

Dicken, P., & Malmberg, A. (2001). Firms in territories: A relational perspective. *Economic Geography, 77*(4), 345–363.

Don Mario. (2005). *Un sueño argentino*. Chacabuco: Don Mario.

Don Mario. (2013). *Una realidad regional*. Chacabuco: Don Mario.

Dunning, J. (2000). The eclectic paradigm as an envelope for economic and business theories of MNE activity. *International Business Review, 9*(1), 163–190.

ECLAC (Economic Commission for Latin America and the Caribbean). (2006). *Foreign investment in Latin America and the Caribbean 2005*. Santiago de Chile: United Nations.

Filomeno, F. (2013). How Argentine farmers overpowered Monsanto: The mobilization of knowledge-users and intellectual property regimes. *Journal of Politics in Latin America, 5*(3), 35–71.

Finchelstein, D. (2012). Políticas públicas, disponibilidad de capital e internacionalización de empresas en América Latina: Los casos de Argentina, Brasil y Chile. *Apuntes: Revista de Ciencias Sociales, 39*(70), 103–134.

Friedmann, H. (1992). Distance and durability. Shaky foundations of the world food economy. *Third World Quaterly, 13*(2), 371–383.

Fuck, M., & Bonacelli, B. (2007). A pesquisa pública e a industria sementeira nos segmentos de sementes de soja e milho hibrido no Brasil. *Revista Basileira de Innovaçao*, 6(1), 87–121.

Fuglie, K., Heisey, P., King, J., Pray, C., Day Rubinstein, K., Schimmelpfennig, D., … Karmarkar-Deshmukh, R. (2011). *Research investments and market structure in the food processing, agricultural input, and biofuel industries worldwide. Economic research report 130.* Washington, DC: USDA.

Gras, C. (2013). *Agronegocios en el Cono Sur. Actores sociales, desigualdades y entrelazamientos transregionales* (Working Paper 50). Berlin: Desigualdades.net.

Gras, C., & Sosa Varrotti, A. (2013). El modelo de negocios de las principales megaempresas agropecuarias. In C. Gras & V. Hernández (Eds.), *El agro como negocio: Producción, sociedad y territorios en la globalización* (pp. 215–236). Buenos Aires: Biblos.

Gudynas, E. (2012). Estado compensador y nuevos extractivismos. *Las ambivalencias del progresismo sudamericano. Nueva Sociedad, 237,* 128–147.

Hubert, B., Goulet, F., Tallon, H., & Huguenin, J. (2013). Agriculture, modèles productifs et options technologiques. Orientations et débats. *Natures, Sciences, Societés, 21,* 71–76.

Idigoras, G. (2013). *Documento de referencia. Mesa Nacional de implementación del NSPE Mejoramiento de Cultivos y Producción de semillas.* Buenos Aires: MINCYT.

INDEC (Argentine Institute of Statistics and Censuses). (2002–2015). *Complejos exportadores.* Retrieved from http://www.indec.gob.ar

Jacobs, E., & Gutierrez, M. (1985). *La industria de las semillas en la Argentina.* Buenos Aires: CISEA.

Kosakoff, B., & Ramos, A. (2010). Tres fases de la internacionalización de las empresas argentinas. Una historia de pioneros, incursiones y fragilidad. *Universia Business Review, 25,* 56–75.

Latour, B. (1994). On technical mediation: Philosophy, sociology, genealogy. *Common Knowledge, 3*(2), 29–64.

Manciana, E., Trucco, M., & Piñeiro, M. (2009). *Large-scale acquisition of land rights for agricultural or natural resource-based use: Argentina.* Buenos Aires: CEO.

McMichael, P. (2009). A food regime genealogy. *The Journal of Peasant Studies, 36*(1), 139–169.

Murmis, M., & Murmis, M. R. (2012). El caso de Argentina. In F Soto Barquero & S. Gómez (Eds.), *Dinámicas del mercado de tierras en América Latina y el Caribe. Concentración y extranjerización* (pp. 15–57). Santiago de Chile: FAO.

Pritchard, B. (2000). *The tangible and intangible spaces of agro-food capital.* Paper presented at the X World Congress on Rural Sociology, International Rural Sociology Association, Rio de Janeiro.

Reca, L. (2012). Agricultura y ganadería en el MERCOSUR. 19080–2010. *Anales de la Academia Nacional de Agronomía y Veterinaria, 66,* 7–22.

Regunaga, M. (2009). *The soybean chain. Implications of the organization of the commodity production and processing industry.* Buenos Aires: Case Studies in Latin America and the Caribbean Region.

Robinson, W. (2015). The transnational state and the BRICS: A global capitalism perspective. *Third World Quarterly, 36*(1), 1–21.

Rocha, P., & Villalobos, V. (2013). *Comparative study of genetically modified and conventional soybean cultivation in Argentina, Brazil, Paraguay and Uruguay.* San José: MAGyP/IICA.

Salles-Filho, S., Carvalho, S., Fuck, M., Libzer, G., Belforti, F., Artunduaga, I., & Vásques, J. (2011). Innovation and intellectual property in the Latin American agricultural sector: An introductory overview for Argentina, Brazil and Colombia. In G. Rozenwurcel, H. Thomas, G. Bezchinsky, & C Gianella (Eds.), *Tecnología + recursos naturales: innovación a escala Mercosur 2.0* (pp. 297–330). Buenos Aires: UNSAM.

Schorr, M., & Wainer, A. (2014). Extranjerización e internacionalización de las burguesías latinoamericanas: el caso argentino. *Perfiles latinoamericanos, 22,* 113–141.

Scoones, I. (2008). Mobilizing against GM crops in India, South Africa and Brazil. *Journal of Agrarian Change, 8*(2–3), 315–344.

SIIA (Integrated System of Agricultural Information, Argentina). (2011). *Superficie sembrada.* Retrieved from http://www.siia.gov.ar

SIP (Subsecretary of Public Incomes, Argentina). (various years). *Recursos tributarios.* Retrieved from http://www.mecon.gov.ar/sip/

Trigo, E., & Cap, E. (2003). The impact of the introduction of transgenic crops in Argentinean agriculture. *Agbioforum, 3*(6), 87–94.

Trigo, E., Cap, E., Malach, V., & Villareal, F. (2009). *The case of zero–tillage technology in Argentina* (Discussion Paper). Washington, DC: IFPRI.

United Nations Conference on Trade and Development. (2005). *World investment report.* New York, NY: United Nations.

United Nations Conference on Trade and Development. (2010). *World investment report.* New York, NY: United Nations.

United Nations Conference on Trade and Development. (2016). *World investment report.* New York, NY: United Nations.

United States Department of Agriculture. (2014). *World agricultural supply and demand estimates 2014.*

Wesz, V. Jr. (2014a). *O mercado da soja e as relaçoes de troca entre produtores rurais e empresas no sudeste de Mato Grosso (Brasil)* (PhD thesis). UFRJ, Rio de Janeiro.

Wesz, V. Jr. (2014b). O Mercado da soja no Brasil e na Argentina. Semelhanças, diferenças e interconexões. *Seculo XXI, 4*(1), 114–161.

Wield, D., Chataway, J., & Bolo, M. (2010). Issues in the political economy of agriculture biotechnology. *Journal of Agrarian Change, 10*(3), 342–366.

Wilkinson, J., & Castelli, P. (2000). *The internationalization of Brazil's seed industry: Biotechnology, patents and biodiversity.* CPDA/UFRJ-Action Aid.

Control grabbing and value-chain agriculture: BRICS, MICs and Bolivia's soy complex

Ben M. McKay

ABSTRACT

This paper analyses Bolivia's industrial value-chain agriculture and argues that a new phase of control grabbing is occurring via value-chain relations. New forms of capital from emerging economies are penetrating Bolivia's countryside and drastically changing the forms and relations of production, property, and power. These processes are analysed by disaggregating the agro-industrial value chain and revealing where the value being produced is appropriated and how the terms of control are changing. The widespread use of genetically modified soybeans and industry's appropriation of natural inputs have opened new spaces for capital to penetrate, circulate, and accumulate, particularly from Brazil, Argentina, and China. As the production process becomes increasingly commodified, smallholders are adversely incorporated into value-chain relations, threatening their ability to work their land now and in the future.

Introduction

Significant agrarian changes are emerging in Bolivia as a result of its insertion into a globalized agro-industrial value chain – itself, part of a much broader global agro-food system. While a convergence of crises around food prices, peak oil, finance, and climate change has fuelled interests in flex crops[1] and commodities, emerging economies such as Brazil, Russia, India, China, and South Africa (BRICS) and some Middle-Income-Countries (MICs) are changing the global political and economic landscape with important implications for global agrarian transformation (McKay, Hall, & Liu, 2016). These changing dynamics are resulting in a 'spatial restructuring process' (McMichael, 2013a) of the global food system, reshaping patterns of production, distribution, and consumption worldwide. Brazil is now a world leader in soybean and sugarcane production, while China's soy imports account for almost two-thirds of the total global soy trade as well as being one of the top five leading soy-producing countries worldwide (ANAPO, 2015; USDA, 2015). The oilseed crop is now being increasingly used and promoted for its potential to be flexed for multiple purposes, predominantly to feed a growing livestock industry in China, but also as edible oil, biodiesel, and bioplastics for industrial uses worldwide (Oliveira & Schneider, 2016).

These converging global dynamics have led to new capital investment, new market opportunities, frontier and industrial expansion, and often forms of dispossession, exclusion, and conflict. While new forms of capital penetration and frontier expansion have been rapidly increasing in Brazil – especially in regard to its agro-industrial sector and the new bioeconomy (McKay & Nehring,

2014) – they are also having broader regional effects among neighbouring countries like Bolivia (Urioste, 2012), Paraguay (Galeano, 2012), and Uruguay. This can be conceptualized of as capital overflow, or what David Harvey calls a spatio-temporal fix, whereby surplus capital seeks to expand and circulate through un- or under-saturated markets as a solution to crises of accumulation. This occurs through frontier expansion, the commodification of land and other natural resources, or the creation of new factors of production – such as agro-industrial inputs – and thus new spaces for capital to circulate. The logic of capital requires constant expansion for accumulation, geographically, and/or sectorally (controlling forward and backward linkages in the supply chain) where convenient and strategic. As Brazil's land market became saturated much earlier, individuals and agribusinesses mainly from Rio Grande do Sul, Parana, and Mato Grosso took advantage of Bolivia's much cheaper and more fertile lands in the 1990s, while agro-inputs from Argentina and China later came to dominate newly established agricultural input markets. This expanding agro-industrial complex is now adversely incorporating and excluding the majority of Bolivia's agricultural producers in the eastern lowlands. This paper analyses these new dynamics of agrarian change by disaggregating the agro-industrial value chain and revealing where the value being generated is appropriated and how the terms of control and access are changing.

Land and control grabbing in Bolivia

This contribution understands contemporary land grabs in Bolivia based on the following three interlinked features put forth by Borras, Franco, Gomez, Kay, and Spoor (2012, pp. 850–851): (i) the power to control land and its productive resources (ie. 'control grabbing'); (ii) large-scale, in terms of either relative land size or capital involved; and (iii) a response to the convergence of multiple crises and the emerging needs for resources by newer hubs of global capital, particularly BRICS and MICs. This characterization of contemporary land grabs delves deeper into land-based social relations of control and access, the multiple dimensions of scale, and the broader changing dynamics of the global political economy. Applying this framework to Bolivia's soy complex enables us to go beyond the land question to broader forms of control associated with value-chain agriculture whereby peasants and smallholders are incorporated into commercial agricultural relations (McMichael, 2013b). Recent studies have highlighted the presence of foreign actors/capital or 'foreignization' of land in Bolivia – particularly focusing on the Brazilian presence in Santa Cruz (Mackey, 2011; Marques Gimenez, 2010; Redo, Millington, & Hindery, 2011; Urioste, 2012).

It has been during the last 20 years that foreigners – specifically Brazilians – have rapidly increased their control over Bolivian agricultural land and resources. In 2006–2007, for example, Brazilians controlled 40.3% of total soybean plantation area in Bolivia, up from 19.6% in 1994–1995 (Urioste, 2012). Although there are no data available on the total amount of land controlled by Brazilians at present, the most reliable and recent study conducted by Miguel Urioste of TIERRA suggests that

> in oilseeds alone, Brazilians own approximately half a million hectares of the best agricultural lands, both category I (intensive agricultural use) and category II (extensive agricultural use), without counting those that are in fallow or rotation, nor those that are directed towards other crops or ranching, which usually comprise larger areas. (2012, p. 449)

Urioste (2012) also suggests that the more recent investments from Brazilians in Bolivia are in pasture lands for cattle ranching. It is estimated that Brazilian cattle ranchers occupy 700,000 hectares in the three provinces bordering Brazil (German Busch, Velasco, and Angel Sandoval) within the

Department of Santa Cruz. Brazilian capital, therefore, controls an estimated 1.2 million hectares of Bolivia's 2.86 million total hectares of cultivated land with Brazilian-based corporations Grupo Monica (Monica Semillas), Gama Group, and UNISOYA controlling over 200,000 ha of land (INE, 2012; Urioste, 2012).

These data from Urioste (2012), however, are based on reports published by Bolivia's politically and economically influential association of large-scale agro-industrialists (*Asociación de Productores de Oleaginosas y Trigo*, ANAPO) which aims to reproduce the Brazilian model of agriculture in Bolivia. ANAPO has access to the most accurate information regarding land tenure (and nationality) since its members report these data to the association. However, ANAPO's publications in recent years no longer include specific information on producer nationality, largely due to publications released by a Bolivian NGO on the issue of foreignization which created a large public backlash not only against ANAPO from its members but also from the public at large, and especially rural worker associations which only recently put the issue against the 'foreignization' of land on their political agenda (Machaca, personal communication, October 2014). It is clear, however, that ANAPO values and encourages foreign investment, especially from Brazil. According to ANAPO's former President, Demetrio Perez, investments from Brazil, Argentina, and other countries have helped and continue to modernize Bolivia's soy sector with new machinery, seed, and agrochemical technologies, expertise, and highway development (Perez, personal communication, February, 2014). Urioste (2012, p. 446) points out that 'two of the leading Brazilian soybean producers serve on the board of the National Association of Soybean Producers (ANAPO), even though this requires changes to organizational statutes'. ANAPO's agenda then is to support the development and expansion of agro-industry for export, representing those medium- and large-scale farmers (22% of total farm units) who control 90% of cultivated soybean area (ANAPO, 2011). Urioste (2012, p. 450) also suggests a general acceptance of the foreign presence – especially among the middle classes of Santa Cruz – so as to secure access to 'sources of capital, technology, employment, business, market knowledge, inputs and genetically-modified seeds'.

This general acceptance is similar to Mackey's (2011) research findings in the region which gives primacy to Brazilian technological transfer in 'manufacturing consent' amongst Bolivian farmers. Like Urioste (2012), Mackey (2011) points to the use of technology as a terrain of legitimation and the informal class alliances among Bolivian and Brazilian agro-industrialists which have led to the 'foreignization' of Bolivia's eastern lowlands. Mackey (2011) also suggests that it is important to consider the Brazilian presence in Bolivia in terms of the much broader politico-economic relationships between the two countries and Brazil's position as a regional hegemony and alternative to Western imperialism. Brazil's role in the production and consumption of Bolivia's hydrocarbon sector, as well its role as a leading creditor, primarily for transportation infrastructure, but also credit for agricultural machinery, has solidified bilateral relations between the countries and led to a general acceptance of Brazilians in the country (Mackey, 2011). According to Brazil's Foreign Affairs Minister for the Coordination of Economic Affairs in South America (*Coordenação-Geral Econômica da América do Sul, Ministério das Relações Exteriores*), Joao Parkinson de Castro, the Ministry always prefers to avoid any discourse regarding 'Brazilians in Bolivia', but it does support Brazilian citizens across the border through political negotiation if necessary. The Minister said that the relationship with the current government is delicate but positive and that they always 'want to avoid any discourses of regional imperialism' (Parkinson de Castro, personal communication, May 2014). He added that

the economic relationship between Santa Cruz and Mato Groso/Mato Groso do Sul is very important, but the politics in La Paz can sometimes threaten this relationship, so it is important that our government supports but does not over-extend its influence in Bolivia. (Parkinson de Castro, personal communication, May 2014)

Colque (2014) reveals the workings of land appropriation on the frontier in Santa Cruz, focusing on the concentration of land-based wealth and power held among agro-industrial elites in Santa Cruz and their ability to challenge and dispute the authority of the state. Colque traces these land-based power relations through several stages of appropriation dating back to the country's first agrarian reform in 1953. Appropriation of land, he argues, predominantly occurs through informal and illegal land deals which exploit and further marginalize the rural poor.

Redo et al. (2011, p. 335) suggest that 'most (of the deforestation) resulted from Brazilian farmers and ranchers moving into the north-east of the region from Mato Grosso do Sul'. These studies, and others, have brought to light the important and contested issue of the foreign (mainly Brazilian) presence in Bolivia's eastern lowlands. It is clear that foreigners, especially Brazilians, have come to control a large share of agricultural land in Bolivia over the past three decades. But what is of interest here is not only if and how foreigners are controlling large parcels of land in Bolivia but how this capital investment is changing relations of production, property, and power; where the value generated by this agro-industrial expansion is appropriated; and the implications and trajectories of this type of agrarian change for the rural majority. Changes in the agrarian political economy, even where foreign capital is present, cannot be fully captured using a 'foreignization' lens. Similarly, an overly land-centric focus can lose sight of relations of debt, dependency, and exclusion whereby land and resource control shift without physical displacement or property rights-based changes.

The focus on nationality presents a veiled threat to understanding the nature, pace, and trajectories of agrarian change in the broader context of new forms and mechanisms of control. The lines distinguishing national from trans-Latina capital have become blurred as joint ventures, subsidiaries, informal partnerships, land leasing, financing, and cross-border marriages render it increasingly difficult to distinguish between Bolivian and Brazilian capital in many instances. Nationality should become central if it leads to capital flight, but not necessarily if it operates similarly to domestic capital. As the national origin of capital and financial investment become intertwined, it becomes more important to reveal the changing relations of access and control over land and resources and the politics behind these processes. This study, therefore, does not analyse the new processes of agrarian restructuring in Bolivia as 'foreignization', but rather in terms of the logic of capital and its incessant drive for accumulation through expanded reproduction, dispossession, and exclusion. Breaking down the imaginary national borders and giving primacy to classes of capital 'distinguished by the interest and strategies of capital in particular activities and sectors and on scales from local to regional, national to transnational' (Bernstein, 2010, p. 112) requires going beyond the foreignization of property to the '*character* and *direction* of change in social relations of property' (Borras et al., 2012, p. 864, emphasis in original).

The development of Bolivia's eastern lowlands: expanding the frontier

Bolivia's first agrarian reform programme of 1953 achieved much success in redistributing landholdings to a large number of households. According to Thiesenhusen (1989, p. 10), 83.4% of the total arable 'forest and agricultural surface' was redistributed to 74.5% of the total number of farming families. This was largely due to the dismantling of the *latifundia*[2] system which restored land back to indigenous groups and peasant farmers. A two-track agricultural development strategy was pursued, distributing parcels of land between 20 and 50 hectares to Andean peasants (*colonizadores*) and 500–50,000 hectare plots to capitalist entrepreneurs and the political and economic elites (Valdivia, 2010, p. 69). This intentional uneven distribution of land is an important historical feature of the development of today's highly unequal agrarian structure and the concentration of capital and wealth that continues to exist in Santa Cruz.

Seeking to open up its agricultural frontier in the fertile lowlands of Santa Cruz, the Bolivian state passed Supreme Decree 4192 in 1955 to allow Mennonites to purchase agricultural land, set up colonies, and develop the new frontier in Bolivia, while guaranteeing their rights to religious freedom and traditional customs. Emigrating mainly from Canada, Mexico, and Belize, Mennonite farmers slowly started to move to Bolivia's eastern lowlands in large groups, purchasing vast rural areas known as 'colonies' (*colonias*) – some up to 20,000 hectares and beyond. With them, they brought capital, machinery, and technological know-how. This phase of expansion overlapped with the Japanese colonization project, who introduced soybeans to Bolivia, based on small-scale production for consumption. Mennonites, however, were the first to introduce, on a much larger scale, highly mechanized agricultural equipment and initiated the first phase of industrial agricultural production in the lowlands of Santa Cruz (Hecht, 2005; Kopp, 2015).

By the 1980s, commercial soy production, driven by Japanese and Mennonite colonists, began to take expand (Hecht, 2005, p. 380). During this period of structural adjustment and the 'eastern-landlord' bias, a massive expansion of soybean plantations emerged by means of deforestation. During the late 1980s and early 1990s, the neoliberal period attracted another phase of foreign investment and frontier expansion. This time, Brazilians and Argentinians took advantage of extremely low land prices. Trade liberalization, financial deregulation, and the World Bank's $56.4 million 'Eastern Lowlands Project' and its Soil Use Plan (*Plan de Uso de Suelos, PLUS*) implemented from 1991 to 1997 facilitated this large-scale land expansion for export-oriented industrial agriculture, namely for soy production (Redo et al., 2011; World Bank, 1998). This project transferred financial and technical resources to large-scale landowners in order to increase their productive capacity for export-oriented development. From 1990 to 1996, agricultural exports from Santa Cruz increased 400%, while the gross value of agricultural output almost doubled from US$ 350 million to US$ 685 million during the same period. Furthermore, transportation linkages were improved with 410 km of road maintenance and improvements (World Bank, 1998). This period attracted many foreigners as land markets opened up and foreign capital, especially from Brazil, penetrated into Bolivia's fertile lowlands.

As one Brazilian producer with 1400 hectares (ha) in San Julian, Bolivia explained, he arrived in Bolivia in 1990 from Rio Grande do Sul (southern-most state of Brazil) because land prices were extremely expensive in the south (of Brazil) and the Bolivian government was trying to attract Brazilian producers to invest in their country and bring their machinery and technological know-how (Klaus, personal communication, April 2014). This Brazilian farmer bought his land for USD $30/ha in 1990s, while its present-day value is estimated between $2000 and $3000/ha. In Brazil, he estimated that the same type of land would cost between $20,000 and $30,000/ha. Moreover, the costs of production are much cheaper in Bolivia, he explained. For example, diesel and gasoline are both subsidized by the state at roughly US$0.50/litre. Using an estimated 40 litres of diesel between sowing and harvesting (Urioste, 2012), this equates to a state subsidy of roughly US$23.6 million per year in mechanized soy production alone.[3] Large-scale industrial soybean plantations also began much earlier in Brazil, so that land became scarce and more expensive, and with technologies still developing for industrial soybean plantations in the Cerrado region, many Brazilian took advantage of Bolivia's easily accessible, cheap, and fertile lands across the border (see Urioste, 2012). Another Brazilian producer, Claudio Batista Vega from Parana, went to Bolivia in 1996 to take advantage of the much cheaper and available land. Batista owns 430 ha, works another 270 ha for farmers with no machinery, and recently bought properties from his neighbours (40 ha each) at a price of USD $1,200/ha. He also owns land in Brazil, but left that for his children to work as he went across the border to expand production. He continues to travel between Bolivia and Brazil based on sowing and harvest cycles, living in both countries throughout the year. Batista explained that he purchased his 430 ha for USD $50/ha and he believes he could sell it now for USD $6000/ha since it is top-quality fertile land producing 3.5 tons/ha which is almost double the region's average. One reason for this is the scarcity of fertile land available for cultivation. Despite claims by Vice President Garcia Linera and agro-industrial representatives of the CAO (*Cámara Agropecuaria del Oriente*) to increase the agricultural frontier by one million hectares per year until 2020 (Heredia Garcia, 2014; Vicepresidente, 2012), it is unclear where this expansion will take place. The soy frontier is already reaching its limits, encroaching into indigenous territories (TCOs) such as Guarayos and Lomerio (Vadillo, Salgado, & Muiba, 2013).

Soybean cultivation in Bolivia has certainly increased rapidly relative to other crops, but neighbouring small countries such as Uruguay and Paraguay have experienced much more rapid expansion, particularly over the past 10 years. This can partly be explained by the limited suitable land available for soybean expansion relative to other countries, but also the politics of the MAS government. Discourses of resource nationalism, food sovereignty, and of an 'agrarian revolution' initially discouraged and cautioned investors from pursuing large-scale land investments. Yet, such popular discourses have failed to result in concrete actions, particularly concerning a viable pathway towards food sovereignty (McKay, Nehring, & Walsh-Dilley, 2014). Rather than large-scale land investments, a new wave of transnational capital is penetrating the countryside not in the pursuit of land per se, but to control the upstream and downstream value-chain components of the soy complex.

Industrial value-chain agriculture and transnational capital

The rapid expansion of soybean plantations, which has doubled in cultivation area since 2000,[4] has transformed productive relations and the development trajectory of the region. Traditional peasant farmers (*colonizadores*) who arrived at Santa Cruz from Potosi, Oruro, Cochabamba, La Paz, among other places, transitioned from their 'peasant way' of cultivating traditional crops such as maize, yucca, rice, and beans, using family labour and producing for household consumption to a capital-intensive model of mechanized agriculture for export. Without access to sufficient capital, many small-scale farmers continued to produce traditional crops until the mid-2000s. Currently, in the two principal communities in the agricultural expansion zone of Santa Cruz – Cuatro Cañadas and San Julián – almost everyone with over 20 hectares of land is engaged in genetically modified (GM) soybean production.

The abandonment of peasant-based farming was triggered by several factors. The first phases of expansion characterized by the arrival of Mennonites followed by Brazilians brought an influx of foreign capital and investment with new machinery and technologies being introduced to the region. As new capital started to penetrate the region in the 1990s, discourses of modernization, progress, and technological advancement via the agro-industrial model also emerged. Transitioning governments, a new agrarian reform programme and highly inefficient land titling process (*saneamiento*) coupled with extremely unfavourable agricultural conditions between 1996 and 2003 (floods, drought, poor yields) and a lack of support for small-scale peasant agriculture resulted in many people leaving their land for opportunities elsewhere. Those who stayed, however, depended on self-exploitation, drudgery, and attempted to diversify their livelihood strategies as much as possible. By the mid- to late-2000s, those who had not already made the transition to soybean cultivation did so, as the economic opportunities of converting one's land from traditional crop production to monocrop soybean production were attractive and offered farmers the chance to 'advance', 'modernize', and obtain a disposable income. This, however, came at the cost of entering into value-chain relations of debt and dependency and for many, the loss of control over their land.

Economic incentives were not the only reason many abandoned diversified crop production for monocultures. Mrs Choque, for example, arrived at the soy expansion region in the late 1980s. The daughter of a peasant farmer and trade union leader of Villa Primavera, her family used to produce maize, rice, yucca, plantains, tomatoes, onions, and other vegetables, while they also had a few heads of cattle and pigs. They were highly autonomous with very little dependence on the market. They worked their land using family labour, producing a surplus after satisfying household consumption needs. In the early 2000s, the Brazilian agribusiness Sojima, which controls over 100,000 ha in the region, purchased vast amounts of land nearby their family's parcel for large-scale GM soybean production. The company uses aircraft fumigation for their crops, contaminating the nearby area with glyphosate and other herbicides and pesticides. As the soil of all nearby parcels became contaminated, they were forced, along with many other peasant farmers, to make the transition to GM (glyphosate resistant) crop production (Choque, personal communication, December 2014). This is a common story not only in her community but throughout the agricultural expansion zone in Santa Cruz.

In her community of Villa Primavera, only 2 out of 20 families have agricultural machinery. The rest, like herself, engage in a rental agreement with someone who has machinery to work their land. She says that a Brazilian landowner works the majority of the land in the community, and though he does not formally own the land, he has direct access to it and derives the most benefits from it. When asked about the future of small farmers in the region, Mrs Choque response was quite grim. 'In the future, small farmers are not going to be able to produce', she says.

> Every year the costs of production are increasing as we need to buy more and more chemicals. The weather has also changed, it is less predictable and we have less rainfall. And since the majority of us (small farmers) don't have access to machinery we are dependent on others and have to wait until they have time to work our land, losing out on the best times for sowing, fumigating, and harvesting.

She went on to explain that in the near future her family plans on diversifying their production system with a variety of vegetables, cattle and pigs, in order to regain the autonomy that they once had. While she worries about the surrounding contamination of the air, land, and water, she was tired of not having any control over her land and the production process. For her, this was not farming as she knows it, but an industrial process exploiting the land and people for profits (Choque, personal communication, December 2014). Despite an increase in household income, the loss of control over the land and the production process, combined with increasing costs for household consumption (i.e. food) is leading to a kind of 'economic upgrading but social downgrading' of smallholders integrated into the soy complex value chain (see Pegler, 2015). Social downgrading in the soy complex is understood here as 'productive exclusion' (see McKay & Colque, 2016), but the social values and peasant identities of present-day smallholders are also being eroded as they become incorporated into value-chain relations and substitute traditional crops for industrial 'flex' crops. Women in particular, like Mrs Choque, have been further downgraded and excluded in the highly mechanized form of production due to their lack of participation in heavy machinery operations (field notes, 2014–2015).

While most smallholders are renting their land or entering into an arrangement *al partida*,[5] processes of productive exclusion are separating farmers from accessing the necessary factors to put their land into production (McKay & Colque, 2016). Others, however, have advanced and built up enough savings to buy a tractor and have become fully integrated into the soy complex, dependent on corporate-controlled agro-industrial inputs such as GM seeds and agro-chemicals (i.e. the technological package complete with growing instructions). Short-term credit and growing contracts have bound farmers into relationships of dependency with agribusiness as they enter into a cycle of indebtedness and control. As McMichael (2013b, p. 671) puts it, 'the producer enters a particular kind of value relation that has the potential to become an instrument of control, debt dependency and dispossession'. This is precisely the type of value relation which has come to control small farmers in Santa Cruz. Farmers' autonomy is threatened, as it becomes nearly impossible to break away from these chains of dependency due to both economic (supplier contracts, indebtedness) and ecological (soil degradation, contamination, large-scale spraying activity) factors.

The introduction of GM soybeans has opened up new market opportunities for agribusiness as Bolivia's untapped agricultural market launched a new frontier of accumulation. Rather than land purchases, transnational capital can still appropriate value from industrial agriculture via agro-inputs, storage and processing facilities, credit and debt relations, and export markets. This is another form of control grabbing in Bolivia's agricultural sector. While existing large-scale land-owners are expanding their landholdings via appropriation and land purchases, and control grabbing via *partida* arrangements (Colque, 2014; McKay & Colque, 2016), transnational agro-capital is penetrating the market via control over seeds, agro-chemical inputs, silos, and export markets. Four of the top six companies which control 85% of soybean market for storage and processing (silos) are owned by foreign capital (AEMP, 2013). With oligopoly control over the soybean market, these six companies are able to set prices and greatly influence crop production. Through supply contracts, these companies have a high degree of control over the production process as they demand specific quality standards which require the use of certain inputs and technological packages. They also have access to export markets and therefore the gateway to where the country's soybeans realize their value. Without actually owning the land or having legal land tenure rights, the relations of control and access to land and its productive resources are largely in the hands of agro-industrial capital. Farmers bear the majority of the risk in this value relation. International price volatility, drought, floods, pests and weeds, and so on are all potential threats that must be absorbed by the producer. Meanwhile, agribusinesses benefit from the sale of agro-inputs – such as seeds, agro-chemicals, machinery, technical assistance, credit – and often bind the buyer of its products (the producer) into selling his or her crops, in their entirety, back to the corporation's silos/processing facilities (field notes, 2014–2015).

The terms of access and control have thus become transformed. Owning land is no longer a sufficient asset when one enters into this particular type of value relation, becoming both dependent on agribusiness for the necessary factors of production and to sell the final product. McMichael (2013b, p. 672), for example, breaks down the value-chain relation as establishing '*chains* of dependency, with smallholders entering markets over which they have no ultimate control', while serving to 'generate *value* that can be appropriated by agribusiness and its financiers – in the commodity form of food, feed and agrofuels for elite consumers, redistributing value from producers to corporate financiers (whether in agribusiness or any other economic sector)'. All the risks of production are, therefore, assumed by the producers, while the value that they add is through labour power and the ecological value extracted from their lands. At the time of harvest, producers sell their crops to the agro-industry, receiving a price per ton which is bound to the Chicago Board of Trade (CBOT) and discounted approximately $70/ton according to the adjustments agreed upon by the six companies which control Bolivia's silo and export markets (ANAPO, 2016; FAOSTAT, 2016). In order to clear and prepare their fields and make the necessary initial investment for the next season,[6] indebtedness through supplier contracts is often a necessity. Debt is therefore a key mechanism of the value-chain relation, 'constituting the "chain" through which such new contract farming is activated, reproduced and, in some cases, dispossessed' (McMichael, 2013b, p. 672). Of course, those with more access to capital, credit, land, and machinery have greater control over the production and decision-making process. Approximately one-third of farmers in Cuatro Cañadas and San Julián own a tractor, meaning that they at least need to hire operational services such as harvesting and transportation or enter into a *partida* arrangement to carry out production from sow to harvest (INE, 2015; field notes 2014–2015). Relations of value-chain control thus vary, as farmers with different access mechanisms are incorporated accordingly.

While it is possible to quantify the value appropriated (or retained) by smallholders in the *partida* arrangement, given that the terms between smallholders and capitalized producers fluctuate between 18% and 25% and the costs of production and soybean prices are available; it is much more difficult to quantify the value appropriated by agribusiness companies in the contract farming schemes. However, we can determine the concentration of control of the upstream and downstream components of the value-chain and the origins of capital and manufacturing. For processors, it is also possible to calculate the crush value of the soybean industry. The soybean crush is the process of converting soybeans into soybean meal and soybean oil and the relationship between their prices is called the Gross Processing Margin (GPM) (CBOT, 2006). The crush spread, or GPM, is a measurement of the profit margin for soybean processors. While many factors affect the soybean crush spread, according to a report by the CBOT, 'soybean prices are typically lowest at harvest and trend higher during the year as storage, interest, and insurance costs accumulate over time' (CBOT, 2006, p. 1). Evidently, this means that producers receive the lowest price for the oilseed grain at harvest, and agribusiness companies who store, process, and trade the crop appropriate more value as the price increases along the chain. Soybeans are crushed into meal (73.3%), oil (18.3%), hulls (6.7%), and waste (1.7%). To calculate the GPM or soybean crush, we use a common denominator of US dollar per metric ton and use the following equation:

$$\text{GPM} = (\text{Price of soybean meal } (403.28) \times 73.3\%) + (\text{Price of soybean oil } (669.86) \times 18.3\%)$$
$$- \text{Price of soybeans } (390.40)$$

Based on prices from the CBOT, the soybean crush spread for July 2016 was USD \$28.15/MT. This means that for every metric ton of soybean produced in Bolivia, agro-industry appropriates USD \$28.15, given the soybean crush spread for July 2016. In the summer harvest of 2014–2015, Bolivia produced 2,106,600 MT of soybeans. Assuming a GPM of USD \$28.15, agro-industry would appropriate USD \$59,300,790 from the crush for that harvest alone. The following table denotes the net revenue gains (or losses) based on data from two periods. Several assumptions are made. First, the hypothetical is based on the *partida* arrangement (25:75), costs of production are based on an average calculated at USD \$463.62/ha, soybean prices are based on Bolivian prices in 2016 and the average during the boom years from 2008 to 2014; yields are assumed at 2 tons/ha; and the GPM is based on CBOT data from July 2016 (field notes, 2014–2015).

As shown in Table 1, agro-industry maintains a consistent revenue based on the GPM rather than directly correlated with soybean prices alone. Producers, who are the capitalized farmers working the land of smallholders, take on most of the risk as they invest in agro-inputs to put land into production often through debt relations, and are thus subject to price and yield volatility and natural disasters. Smallholders forfeit the use of their land and may benefit

Table 1. Revenue distribution of the soy complex: smallholders, producers and agro-industry.[9]

	Partida contract	Costs of Production ($USD/ha)	Soybean Price	Yield (ton/ha)	Gross Revenue/ ha	Net Revenue/ ha	Partida/ ha	Partida/ ton
Net revenue for soybean production, 2016 prices								
Smallholder	25% of net revenue	463.62	230	2	460	−3.62	0	0
Producer	75% of net revenue	463.62	230	2	460	−3.62	(−)3.62 X ha	net loss
Agro-industry								28.15
Net revenue for soybean production, average prices from boom years (2008–2014)								
Smallholder	25% of net revenue	463.62	338.88	2	677.76	214.14	53.54	26.7675
Producer	75% of net revenue	463.62	338.88	2	677.76	214.14	160.61	80.3025
Agro-industry								28.15

minimally (as during the boom years from 2008 to 2014) or may receive nothing (as in 2016) but avoid entering into debt. During boom years, producers did retain significant profits of approximately USD $160/ha when working the land of smallholders in a *partida* arrangement. When working their own land with access to all means of production (tractor, harvester, and fumigator) and thus able to avoid operational costs such as machinery rentals, producers retain profits upwards of $207/ha based on 2016 figures and $425/ha during the boom years. This is the ideal type of producer profile which all smallholders aspire and which pulls farmers into the soy complex. However, those who retain such profits represent between 5% and 20% of the total population of soybean producers. Most smallholders are increasingly marginalized, forced to sell their labour power, as income derived from the *partida* arrangement is uncertain, subject to volatile prices, unpredictable climate conditions and inconsistent yields. Agro-industry, on the other hand, not only appropriates surplus value from the soybean crush, it also has transformed natural components of the agricultural production process into industrial activities to be re-incorporated as agricultural inputs, appropriating value from these new agro-industrial inputs (Goodman, Sorj, & Wilkinson, 1987).

Capital penetration via industrial value-chain agriculture has managed to 'create sectors of accumulation by re-structuring the inherited "pre-industrial" rural production process' (Goodman et al., 1987, p. 8). This occurs through what Goodman et al. (1987, p. 2) call 'appropriationism', defined as 'the discontinuous but persistent undermining of discrete elements of the agricultural production process, their transformation into industrial activities, and their re-incorporation into agriculture as inputs'. As accumulation processes become limited in the agricultural sector due to inherent natural plant cycles, industrial capital seeks to *appropriate* any and all factors of production including seeds, organic inputs, labour, and land. This has occurred in Bolivia's lowlands with GM seeds, agro-chemicals, agricultural machinery, and land markets. At the other end of the value-chain, agricultural crops are increasingly being substituted or flexed as an industrial input – what Goodman et al. (1987) have termed 'substitutionism'. This is even more evident today, as crops such as corn (Gillon, 2016); oil palm (Alonso-Fradejas, Liu, Salerno, & Xu, 2016); soybeans (Oliveira & Schneider, 2016); and sugarcane (McKay, Sauer, Richardson, & Herre, 2016) are increasingly used in multiple ways (food, animal feed, fuel, industrial material) and can be (or thought to be) 'flexed' according to market conditions (Borras, Franco, Isakson, Levidow, & Vervest, 2016). Through scientific and technological advancement, industrial value-chain agriculture has thus appropriated and substituted the natural inputs and outputs of farming to render it as industrial as possible and open new possibilities for commodification and capital accumulation.

Appropriationism and the technological packages complete with seeds, agro-chemicals, and application instructions it has facilitated have led to increases in both costs and quantity of inputs

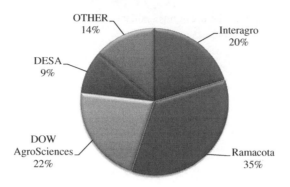

Figure 1. Soybean seed import market 2014 (99.8% from Argentina).[7]

used in production. In 2004, for example, Bolivia imported 198 tons of soybean seeds at an average cost of US$301/ton; in 2012, seed imports amounted to 9862 tons, an increase of 4881%, with an average cost of US$738/ton (AEMP, 2013; INE, 2012). During the same period, soybean cultivation area increased from 852,000 ha to 1,103,390 ha (29.5%). This exponential increase of seed imports which vastly outpaces area expansion is largely due to the legalization of GM soybean seeds. In 2005, GM seeds from Argentina came to dominate the market, accounting for an average of 99.9% of Bolivia's soybean imports from 2005 to 2015 (INIAF, 2005–2014). Although this has led to a proliferation of agro-chemical and GM seed distributors, just four companies control 86% of Bolivia's GM seed distribution market (see Figure 1).

Since GM soybean seeds are engineered to tolerate the herbicide glyphosate, it comes as no surprise that a positive correlation exists between the increase in both the use of GM soybean seeds and herbicides, not only in Bolivia but also throughout the region (Catacora-Vargas et al., 2012). Furthermore, the increased use of glyphosate combined with the adoption of a no-tillage seeding system has resulted in the 'appearance of weeds resistant to glyphosate in GM soybean production … resulting in greater application of complementary herbicides' (Catacora-Vargas et al., 2012, p. 32). Any farmer in the soy expansion zone can attest to this. Mr Fehr, a Mennonite who came to the region in the early 1980s from Mexico to work the land, explains:

> Before the agro-chemicals were better, one chemical took care of everything. Now the technological packages require one product for one pest, another for a different pest, another for a weed and so on. Almost every year we have a new type of weed or pest that must be dealt with. Costs are increasing, but our yields are not. The only ones who keep benefiting are the agribusiness companies selling the chemicals. (Fehr, personal communication, January 2015)

In San Julián, Marcos Churquina Cabezas, the president the Small Producer's Association of the Orient (APPAO) who arrived at the region in 1982 from Potosi, has a similar assessment of GM seeds and agrochemical use. 'In the 1980s', he said, 'we worked with our own labour, no agro-chemicals, no heavy machinery. Now, there are more pests and bad weeds and you cannot harvest anything without agro-chemicals, everything has changed' (Churquina Cabezas, personal communication, November 2014). Churquina Cabezas explained that many farmers transitioned to GM soybean production due to contamination caused by fumigator planes which affected neighbouring land, destroying crops not resistant to glyphosate and other agro-chemicals.

Roberto Churata, the founder of the Agricultural Chamber for Small Producers of the Orient (CAPPO), also shared his insights regarding GM seeds and agro-chemical inputs for small farmers:

> Many farmers today cannot even name the agro-chemical they used last year, or what pests and weeds are affecting their crops. The separation of their relationship with the land has rendered them more individualistic and yield-focused. Of course, many do not know the science behind GM seeds and are not aware that agro-chemicals remain in the soil for many years after application. They are caught up in the vicious circle of dependence on technology, increasing costs, and more pests and weeds. It is almost impossible for them to escape this technological trap since they are thinking on a day-to-day or harvest-to-harvest basis. Questions about sustainability and land for future generations come secondary and will be dealt with down the road. (Churata, personal communication, January 2015)

Churata, who in 2015 was a Senior Official for the mayor's office in San Julián and also works as a private consultant in rural development, is one of the most important political figures representing the interests of small farmers in Santa Cruz. He remains an important advisor and affiliate of the region's largest small farmer associations (APPAO and ACIPAC) with a deep understanding of the politics and socio-economic changes taking place regarding the soy complex.

Another key informant who works as an agricultural engineer and agronomist for ANAPO in San Julián also said that pests and plant diseases are increasing every year which require more and more agro-chemicals. This agronomist works with over 1000 farmers in the region, making daily visits to multiple farms each day. Working for ANAPO for nearly 10 years, he has seen first-hand the changes taking place on the land, as pests and weeds which never existed in the region are suddenly appearing. When asked about the future of smallholders given the current trajectory, this informant's perspective was regrettably grim.

> Small farmers with less than 50 hectares and no or little machinery won't survive in this region; they will eventually be bought out, go into debt and be forced to sell their land. It's difficult to say, but I have seen this process unfold over the past 7 years. (personal communication, January 2015)

Similar stories and farmer testimonies could be mentioned, but data on agro-chemical imports are also quite telling.

According to Bolivia's National Service for Agricultural Health and Food Safety (*Servicio Nacional de Sanidad Agropecuaria e Inocuidad Alimentaria, SENASAG*), from 2010 to 2014, the quantity of agro-chemicals registered in Bolivia increased from 12.6 million kg/l to 38.3 million kg/l in 2014 – a 204% increase, while area under cultivation increased by just 28% (ANAPO, 2015; SENASAG, 2014). Based on both quantitative and qualitative data, it is clear that since the introduction of GM soybean seeds, agro-chemical consumption has increased at rates much higher than cultivation area increases. Furthermore, Figure 2 shows the origin of these agro-chemicals over the same time period, with China, Argentina, Brazil, and Paraguay accounting for 84% of Bolivia's agro-chemical market.

The development of industrial agriculture through value-chain relations opens up new frontiers for capital to circulate and accumulate. But what we see in Bolivia is the importation of agro-inputs (seeds, agro-chemicals, machinery) and the exportation of a raw or semi-processed agro-'outputs' (soybeans and derivatives). With both ends of this value-chain largely controlled by foreign capital, the soybean complex in Bolivia essentially extracts ecological value from its fertile lands, while the value-added activity (surplus value generated) is appropriated elsewhere. Due to its highly mechanized character, the need for labour is also diminished, resulting in diminished employment opportunities for capital-poor and landless rural people (McKay & Colque, 2016). The extractivist nature of this type of agro-industrial development – which we may call 'agrarian extractivism' – parallels the non-renewable resource-extractive economy (minerals, natural gas) which has characterized Bolivia for the past 500 years (Alonso-Fradejas, 2015; McKay, 2017). While similarities can be drawn between the extraction of the ecological value (soil, water, air pollution) and the extraction of the surplus value generated from

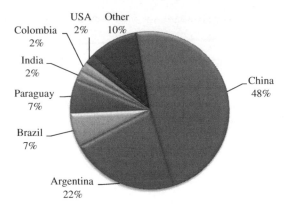

Figure 2. Origin of agro-chemicals in Bolivia, 2009–2014.[8]

exporting a raw or semi-processed resource (minerals, natural gas, soybeans and derivatives), such lines cannot be drawn with the issue of labour. Unlike labour-intensive crop production such as sugar-cane, soybean production is actually excluding the majority of Bolivia's farmers from engaging in productive activity, potentially leading to 'surplus populations' and a 'truncated trajectory of agrarian transition' through processes of dispossession (Li, 2011; McKay & Colque, 2016).

Control grabbing and the spatio-temporal fix

Value-chain agriculture has generated a new phase of control grabbing through particular social relations of production which enable agro-industrial capital to control land and its productive resources without necessarily having tenure rights to the land. This is due to capital-intensive access mechanisms facilitated by the appropriation and commodification of the means of production and value-chain relations of debt and dependency. Value-chain agriculture has created a spatio-temporal fix whereby surplus capital is able to circulate and accumulate, appropriating surplus value and later exported as a raw material for further value-added processing elsewhere. Harvey (2003, p. 115) explains the spatio-temporal fix as 'a particular kind of solution to capitalist crises through temporal deferral and geographical expansion'. What the spatio-temporal fix requires is

> the production of space, the organization of wholly new territorial divisions of labour, the opening up of new and cheaper resource complexes, of new regions as dynamic spaces of capital accumulation, and the penetration of pre-existing social formations of capitalist social relations and institutional arrangement … (which) … provide important ways to absorb capital and labour surpluses. (Harvey, 2003, p. 116)

As agro-industrial soybean production developed much earlier in Argentina and Brazil – both growing hubs of global (agro)-capital – Bolivia offered both a strategic and convenient space to absorb capital surpluses. In the 1990s, for example, when many Brazilians purchased land in Bolivia, the Brazilian land market was becoming saturated, expensive, and newer technologies were still developing in order to expand into the *Cerrado* region (Marques Gimenez, 2010; Urioste, 2012). This phase of investment was prompted by the opening up of Bolivia's land markets in Santa Cruz, offering new and cheaper spaces for capital absorption. Medium- and large-scale Brazilian landowners such as Iglenio Klaus and Claudio Batista Vega were among those seeking new 'green-field sites' for expansion as land markets in the south of Brazil were inflated and saturated. While the resettlement programmes in the 1980s were engineered to provide a 'fix' for labour surpluses after neoliberal policies and the tin price crisis generated widespread employment among the miners, the opening up of the fertile lowlands of Santa Cruz in the 1990s absorbed surplus agro-capital, particularly from Brazil and Argentina.

With land markets at a point of near saturation and scarcity, capital has penetrated once again via value-chain technologies, appropriationism, and debt relations. Capital has managed to penetrate peasant farming, transforming peasants into small-scale capitalist producers, rentier capitalists, and wage labourers. This continues to drastically change the social relations of production, power, and property in this region. As Harvey (2003, p. 116) states, 'such geographical expansions, reorganizations, and reconstructions often threaten, however, the values already fixed in place (embedded in the land) but not yet realized'. Instead of producing crops for household and local consumption, for example, producers now purchase increasingly expensive, external agro-inputs controlled and produced by foreign capital and after adding labour and ecological value, sell this product to a monopolized market controlled by foreign capital for export. As a spatio-temporal fix for foreign capital, value-chain agriculture 'encompasses smallholder farms as "resource complexes" to absorb and create capital' (McMichael, 2013b, p. 674).

Since Bolivia does not have the capacity to absorb the surplus value created in its domestic market, it is used as a space to temporarily absorb capital and add (mainly) ecological value, while China – which imports almost two-thirds of the global soy trade – absorbs (indirectly) the surplus value created on a global scale. This is the temporal aspect of the spatio-temporal fix. Very little of the value-added components of the soy complex are absorbed in Bolivia; capital temporarily penetrates the countryside, circulates through the soil and is exported in its commodity form as a soybean to external markets where it is further processed and fed into the global grain–feed–meat complex. Taking a broader perspective on value appropriation, we can observe that China largely benefits from both ends of industrial value-chain agriculture for soybean production. First, as a producer of agro-chemicals – a processed, value-added product exported around the world; and second, as the world's largest importer and processor of soybeans primarily to feed a growing grain–feed–meat complex. In order to appropriate more value-added economic activity domestically, China adopted an import strategy through a differential import tax structure in 1998 which encourages whole soybean imports (3% import tariff) over soy meal (5%) and oil (9%) (Lee, Tran, Hansen, & Ash, 2016). This import strategy enables processing to occur within China, facilitating industrial value-added production in-country. Furthermore, since China's 'going out' (*zou chuqu*) strategy, its political and economic influence in Bolivia and Latin America more generally has increased substantially as countries move away from the Washington Consensus to a seemingly less stringent 'Beijing Consensus'. However, it remains doubtful that such 'south–south' relations will lead to any meaningful structural changes in terms of altering the region's dependence on raw material exports (McKay, Alonso-Fradejas, Brent, Xu, & Sauer, 2016).

Conclusion

This paper has attempted to disaggregate Bolivia's value-chain agriculture and demonstrate how a new phase of control grabbing has emerged via the penetration of agro-capital, particularly from Brazil, China, and Argentina. This type of control grabbing does not necessarily entail having tenure rights to the land, but rather having control over land-based resources via a value relation characterized by debt and dependency. This has been conceptualized as a spatio-temporal fix, triggered first by Brazil's saturated land market in the south which brought many Brazilian agro-capitalists to Bolivia's lowlands, and second by the commodities boom in the 2000s and the subsequent crises (food prices, financial, and climate) which set in motion increased global investments in land and natural resources.

The introduction of GM soybean seeds facilitated a new phase of capital penetration into Bolivian agriculture as industry appropriated farming's natural inputs and developed new value-chain relations of production, reproduction, property, and power. The vast increase in quantities and prices of agro-chemicals has weakened producers' position vis-à-vis agribusiness, strengthening the relations of debt and dependency. Rather than a form of industrial agricultural development which implies value-added processing, sectoral linkages and employment generation, Bolivia's soy complex is better characterized as 'agrarian extractivism' due to the very social, economic, and environmental extractive dynamics of production (McKay, 2017). Capital overflow from Argentina and Brazil, two of the regions' most advanced agro-industrial hubs and largest economies, continues, in different ways, to penetrate the Bolivian countryside. Yet, rather than absorbing this capital over-flow, the Bolivian soil is used as a spatial fix in which capital temporarily penetrates and circulates, only to be exported for value-added processing elsewhere. The extractive character of Bolivia's soy complex is, therefore, leading to a reprimarization of the Bolivian economy towards even greater dependence on natural resource extraction for primary commodity exports – this time in the form of soybeans rather than natural gas or minerals.

This paper concludes that control grabbing continues in Bolivia – primarily through resource control by means of value-chain relations – appropriating and concentrating value in the hands of transnational corporations. While dominant discourses maintain that this model of rural development is generating employment, contributing to food security and food sovereignty, reducing costs of production, and benefitting small-scale farmers, this paper has argued otherwise. The current trajectory of agrarian change is threatening the ability of small-scale farmers to work their lands, increasing the country's dependency on food imports and thus volatile international markets, and leading to very extractive social, economic, and environmental dynamics.

Notes

1. 'Crops that have multiple uses (food, feed, fuel, industrial material) that can be easily and flexibly inter-changed' (Borras et al., 2012, p. 851). See also TNI Flex Crop Working Paper Series.
2. Large-scale landed estates.
3. Based on the combined area of soybean cultivation in summer 2013 (1,180,000 ha). Author's own calculation based on data from ANAPO (2013).
4. From 490,000 hectares in 2000 to 942,000 hectares in 2014 (ANAPO, 2014).
5. The 'partida' arrangement is a form of land leasing that was not practised before the soybean 'boom', but has now become common in the lowlands where land is relatively scarce. 'Partida' or 'al partir' means to share or split harvest or usufruct benefits among those working the land and those who hold tenure rights to the land (McKay & Colque, 2016).
6. Production costs range from US$420 to US$560 per hectare in Bolivia's expansion and integrated zones (IBCE 2014).
7. Author's elaboration based on data received from personnel at the Instituto Nacional de Innovación de Agropecuaria y Forestal (INIAF) in 2014.
8. Author's elaboration based on data from SENASAG (2014).
9. Author's own based on field notes 2014–2015.

Acknowledgements

An earlier version of this paper was presented at the BRICS Initiative for Critical Agrarian Studies (BICAS) International Conference held at the Institute for Poverty, Land and Agrarian Studies

(PLAAS), University of Western Cape, 20–23 April 2015. I would like to thank the participants at that conference, two anonymous reviewers, and the guest editors of this special issue for their very helpful comments and criticisms. Jun Borras, Max Spoor, Murat Arsel, and Christina Schiavoni also provided valuable feedback at a seminar held at the International Institute of Social Studies (ISS) for which I am grateful. Any remaining errors are, of course, my own.

Disclosure statement

No potential conflict of interest was reported by the authors.

Funding

This work was supported by the Social Sciences and Humanities Research Council of Canada [752-2012-1258] and a BICAS Small Grant Award.

References

AEMP. (2013). *Estudio Mercado Del Grano de Soya*. La Paz: Autoridad de Fiscalizacion y Control Social de Empresas.

Alonso-Fradejas, A. (2015). Anything but a story foretold: Multiple politics of resistance to the agrarian extractivist project in Guatemala. *The Journal of Peasant Studies*, *42*(3–4), 489–515.

Alonso-Fradejas, A., Liu, J., Salerno, T., & Xu, Y. (2016). Inquiring into the political economy of oil palm as a global flex crop. *The Journal of Peasant Studies*, *43*(1), 141–165.

ANAPO. (2011). *Memoria Anual 2011*. Santa Cruz: Asociación Nacional de Productores de Oleaginosas y Trigo.

ANAPO. (2014). *Informe de Campaña Vareno 2013-2014: Soya*. Santa Cruz: Author.

ANAPO. (2015). *Memoira anual 2015*. Santa Cruz: Author.

ANAPO. (2016). *Soya sube de precio en la Bolsa de Chicago*. Retrieved from http://www.anapobolivia.org/noticias.php?op=1&tipo=1&id=649

Bernstein, H. (2010). *Class dynamics of agrarian change*. Halifax and Winnipeg: Fernwood.

Borras, S. M. J., Franco, J. C., Gomez, S., Kay, C., & Spoor, M. (2012). Land grabbing in Latin America and the Caribbean. *The Journal of Peasant Studies*, *39*(3–4), 845–872.

Borras, S. M. J., Franco, J. C., Isakson, S. R., Levidow, L., & Vervest, P. (2016). The rise of flex crops and commodities: Implications for research. *The Journal of Peasant Studies*, *43*(1), 93–115.

Catacora-Vargas, G., Galeano, P., Agapito-Tenfen, S. Z., Aranda, D., Palau, T., & Onofre Nodari, R. (2012). *Soybean production in the southern cone of the Americas: Update on land and pesticide use*. Tromso: GenOk–Centre for Biosafety.

CBOT. (2006). *CBOT soybean crush*. Chicago, IL: Author.

Colque, G. (2014). *Expanión de la frontera agrícola: Luchas por el control y apropiación de la tierraen el oriente boliviano*. La Paz: Fundación TIERRA.

FAOSTAT. (2016). *Food balance*. Retrieved from http://faostat3.fao.org/browse/FB/FBS/E

Galeano, L. A. (2012). Paraguay and the expansion of Brazilian and Argentinian agribusiness frontiers. *Canadian Journal of Development Studies/Revue canadienne d'études du développement, 33*(4), 458–470.

Gillon, S. (2016). Flexible for whom? Flex crops, crises, fixes and the politics of exchanging use values in US corn production. *The Journal of Peasant Studies, 43*(1), 117–139.

Goodman, D., Sorj, B., & Wilkinson, J. (1987). *From farming to biotechnology: A theory of agro-industrial development.* Oxford: Basil Blackwell.

Harvey, D. (2003). *The new imperialism.* Oxford: Oxford University Press.

Hecht, S. B. (2005). Soybeans, development and conservation on the Amazon frontier. *Development and Change, 36*(2), 375–404.

Heredia Garcia, H. (2014). *Gobierno so abre a construer la agenda del millon de ha.* Retrieved from http://www.eldeber.com.bo/economia/gobierno-abre-construir-agenda-del.html

INE. (2012). *Información Estadística.* La Paz: Author.

INE. (2015). *Primer Censo Agropecuario 2013.* La Paz: Author.

Kopp, A. J. (2015). *Las colonias menonitas en Bolivia: antecedentes, asentamientos y propuestas para un diálogo.* La Paz: Fundación TIERRA.

Lee, S. T., Tran, A., Hansen, J., & Ash, M. (2016). *Major factors affecting global Soybean and products trade projections.* United States Department of Agriculture-Economic Research Service. Retrieved from: https://www.ers.usda.gov/amber-waves/2016/may/major-factors-affecting-global-soybean-and-products-trade-projections/.

Li, T. M. (2011). Centering labor in the land grab debate. *Journal of Peasant Studies, 38*(2), 281–298.

Mackey, L. (2011). *Legitimating foreignization in Bolivia: Brazilian agriculture and the relations of conflict and consent in Santa Cruz, Bolivia.* Paper presented at the international conference on global land grabbing, University of Sussex, UK.

Marques Gimenez, H. (2010). *O desenvolvimento da cadeia produtiva da soja na Bolívia e a presença brasileira : uma história comum.* Universidade de São Paulo.

McKay, B., & Colque, G. (2016). Bolivia's soy complex: The development of 'productive exclusion'. *The Journal of Peasant Studies, 43*(2), 583–610.

McKay, B. M. (2017). Agrarian extractivism in Bolivia. *World Development, 97*, 199–211.

McKay, B. M., Alonso-Fradejas, A., Brent, Z., Xu, Y., & Sauer, S. (2016). China and Latin America: Towards a new 'consensus' of resource control? *Third World Thematics: A TWQ Journal, 1*(5), 592–611.

McKay, B. M., Hall, R., & Liu, J. (2016). The rise of BRICS: Implications for global agrarian transformation. *Third World Thematics: A TWQ Journal, 1*(5), 581–591.

McKay, B., & Nehring, R. (2014). *Sustainable agriculture: An assessment of Brazil's family farm programmes in scaling up agroecological food production* (IPC-UNDP Working Paper, (123)).

McKay, B., Nehring, R., & Walsh-Dilley, M. (2014). The 'state' of food sovereignty in Latin America: Political projects and alternative pathways in Venezuela, Ecuador, and Bolivia. *The Journal of Peasant Studies, 41*(6), 1175–1200.

McKay, B., Sauer, S., Richardson, B., & Herre, R. (2016). The political economy of sugarcane flexing: Initial insights from Brazil, Southern Africa and Cambodia. *The Journal of Peasant Studies, 43*(1), 195–223.

McMichael, P. (2013a). Land grabbing as security mercantilism in international relations. *Globalizations, 10*(1), 47–64.

McMichael, P. (2013b). Value-chain agriculture and debt relations: Contradictory outcomes. *Third World Quarterly, 34*(4), 671–690.

Oliveira, G. de. L. T., & Schneider, M. (2016). The politics of flexing soybeans: China, Brazil and global agroindustrial restructuring. *The Journal of Peasant Studies, 43*(1), 167–194.

Pegler, L. (2015). Peasant inclusion in global value chains: Economic upgrading but social downgrading in labour processes? *The Journal of Peasant Studies, 42*(5), 929–956.

Redo, D., Millington, A. C., & Hindery, D. (2011). Deforestation dynamics and policy changes in Bolivia's postneoliberal era. *Land Use Policy, 28*(1), 227–241.

SENASAG. (2014). Database: Servicio Nacional de Sanidad Agropecuaria e Inocuidad Alimentaria.

Thiesenhusen, W. C. (1989). *Searching for Agrarian reform in Latin America.* London: Unwyn Hyman.

Urioste, M. (2012). Concentration and 'foreignisation' of land in Bolivia. *Canadian Journal of Development Studies/Revue Canadienne D'études Du Développement, 33*(4), 439–457.

USDA. (2015). *World agricultural supply and demand estimates*. Author, pp. 1–40. http://doi.org/WASDE-525

Vadillo, A., Salgado, J. M., & Muiba, S. N. (2013). *Gobernanza de los recursos naturales en Lomerío*. La Paz: Fundacion TIERRA.

Valdivia, G. (2010). Agrarian capitalism and struggles over hegemony in the Bolivian lowlands. *Latin American Perspectives*, *37*(4), 67–87.

Vicepresidente. (2012). *Gobierno y CAO acuerdan ampliar la frontera Agricola para garantizar soberanía alimentaria*. Retrieved from http://www.vicepresidencia.gob.bo/Gobierno-y-CAO-acuerdan-ampliar-la

World Bank. (1998). *Eastern lowlands: Natural resource management and agricultural production project*. Washington, DC: Author.

The agrifood question and rural development dynamics in Brazil and China: towards a protective 'countermovement'

Fabiano Escher, Sergio Schneider and Jingzhong Ye

ABSTRACT

This paper explores some features of the development paths taken by Brazil and China (two member countries of the BRICS grouping) in the current context of the crisis of neoliberal globalization and transformation of the political and economic world order. The authors use Polanyi's 'double movement' thesis to argue that newly emerging rural development (RD) dynamics in China and Brazil are part of a protective 'countermovement', driven by actors and institutions responding to the contradictions of the concentration and internationalization of agrifood systems. However, the *direction* and scope of these countermovements are still open; their transformative potential should be viewed in Gramsci's terms as a struggle for hegemony the outcome of which depends on the concrete 'balance of social forces'. First, the paper characterizes the impacts of China's rise on Brazil's development, which subsequently found its economy under threat of reprimarization and deindustrialization. The paper then sketches some stylized facts of production and consumption within the Brazil–China soy–meat complex, a key element of the current global food regime, with a focus on corporate control of the soy–meat value chain, and its negative consequences. Finally, the paper identifies the key roles that actors and institutions linked to peasants and family farmers are playing in the RD dynamics of each country. Although China and Brazil represent two very different realities, the comparison shows that critical rural and agrifood issues are indeed moving onto the centre stage of the contemporary 'double movement'.

Introduction

It is widely acknowledged that the rise of China as a great power has had enormous impacts on the contemporary economic and political world order. Especially, since the 2008 financial crisis unfolded into a 'great recession', a new context seems to be emerging, one in which some countries from the Global South are leaving behind their historically peripheral condition and beginning to play a more significant role in the dynamics of global capitalism. This has led to increasing academic and social interest in the recent trajectories of countries expected to become the new economic powers of the twenty-first century, including those in the so-called BRICS grouping (Brazil, Russia, India, China, and South Africa).

A number of authors have suggested the value of returning to the works of Karl Polanyi (1977, 2000) to understand these developments. His most famous book, *The Great Transformation*

(Polanyi, 2000), analysed the rise and fall of the late nineteenth century liberal world order in which the whole economy and society were subordinated to a system of self-regulating markets built upon the institution of the gold standard as well as British hegemony. Recent analyses point out the relevance of considering the critical moments of neoliberal globalization and US hegemony as a contemporary rerun of Polanyi's 'great transformation' (Arrighi & Silver, 2003; Burawoy, 2003; Evans, 2008; Levien & Paret, 2013; Schneider & Escher, 2011; Somers & Block, 2005). Central to these arguments is the Polanyian thesis of a 'double movement' inherent in the contradictory dynamics of capitalism. On the one hand, there is a movement of disembedding the markets from public regulation and social controls through legislative and institutional changes that commoditize labour, land, and money (which Polanyi calls 'fictitious commodities'), leading to destructive consequences and threats to the livelihoods of people, the natural environment and resources, as well as changes in the organization of economic activities.

On the other hand, there is a countermovement for social and economic protection of those affected by the commodification process. Here social actors engage in collective actions to struggle against the assaults of this 'satanic mill' and advocate for public intervention to circumscribe the power of markets, perhaps leading to the re-embedding of the substantive economy. However, all these authors state, in a more or less explicit manner, that the countermovement should not be viewed as a spontaneous and semi-automatic reaction to the contradictions of self-regulating markets, but as a dialectical process involving power relations and institutional mediations. From this angle, the actors, meanings, and potential outcomes of the countermovements must be evaluated, according to Gramsci's (2002) view, as a struggle for hegemony, since they depend on realization of the concrete 'correlation of forces'.

In the international literature on agrarian political economy and development studies, some authors have placed at the centre stage of the contemporary double movement the relationship between the globalization of developing countries' agrifood systems, led by large agribusiness transnational corporations, or 'food empires' (Friedmann, 2009; McMichael, 2013; Ploeg, 2008; Wilkinson, 2009), and new rural development (RD) dynamics emerging in response to its deleterious effects (Hebinck, Ploeg, & Schneider, 2015; Ploeg, Ye, & Schneider, 2012). In relation to Brazil and China, as well as several European countries, it has been argued that the new RD dynamics currently under way involve the re-alignment of agriculture to nature and society to create new foundations for food production, distribution, and consumption. This represents a counter-hegemonic expression of resistance, resilience, and autonomy of peasants and family farmers, supported by the organizations of strategic political allies (consumers, intellectuals, and policy-makers). In this view, RD is seen as part of a 'countermovement' of actors and institutions in response and reaction to what economists call 'market failure' – in other words, the contradictions and consequences of liberalized and globalized agrifood systems (Ploeg, 2011; Ploeg, Ye, & Schneider, 2010, 2012; Ventura & Ploeg, 2010).

Why are discussions on food and RD taking place in Brazil and China at the same time? How are the relations between these countries shaping the transformations of their agrifood systems? To what extent does the role played by different social actors and institutions in the new RD dynamics currently emerging in both countries allow us to interpret this as part of a Polanyian 'countermovement'? The authors discuss these questions in both countries over the same period – from the mid-1990s, with the consolidation of a market economy in China after the Tiananmen Square protests, and the consolidation of neoliberalism in Brazil after the victory of Fernando Henrique Cardoso, to the present. In the mid-1990s, the effects of globalization began to emerge clearly in both economies, laying the bases for subsequent internationalization of key actors in the agrifood systems of Brazil and China. This is also the point at which new rural livelihood strategies, social practices, and organizational initiatives began to emerge from peasants and farmers themselves, and the point at which new specialized state institutions and policies came into being.

Burawoy (2003) offers a framework for institutional analysis derived from a synthesis of Gramsci and Polanyi. This approach might be useful for examining how key actors engage with each other and the institutions that surround them to produce structural changes across the state, markets, and society. From this perspective, the concept of 'society' assumes both historical specificity and theoretical generality, as it places actors and institutions at the core of the balance of social forces in the struggle for hegemony. While 'civil society' (Gramsci) or 'society as a whole' (Polanyi) reflects the crises and contradictions of capitalism, it is also the terrain of its transcendence. Any process of progressive social change and development triggered by a countermovement necessarily requires interactions between the social actors (society) mediated by the institutions constituted in the spheres of politics (states) and economy (markets). As the historical process evolves, the habitual attitudes of the actors change in response to the changes in institutions, and the determinants of the existing social structures are fully reproduced, partially modified, or radically transformed.

Although Polanyi (2000) had a limited understanding of capitalist hegemony based in the realm of production when compared to Gramsci (2002), he was nonetheless able to understand that the awareness of broader societal interests has its *focus* in the lived experiences shared by people, and that the *locus* of transformative potential resides in the realm of circulation, in the perception that turning livelihoods, public services, and social rights into commodities is destructive. The politicization of circulation and consumption thus provides transformational potential to the universal experience of markets under capitalism, and creates the social foundations for counter-hegemonic movements. Avoiding economic reductionism, progressive social change is seen as a political project, which further depends on the realization of actors' ideas, practices, and struggles to subordinate the economy to 'self-regulating society'. However, this is not to say that the politics of production does not matter. Rather, the point is that the realms of circulation and consumption *also* matter. Concretely, both realms represent two distinct political 'moments' of the hegemonic struggles, and one can strengthen the other.

This paper is structured into four sections. The introduction provides the background. The second section briefly assesses the economic challenges that the rise of China poses for Brazil, given recent trends of trade and investment. The third section, subdivided in two, firstly provides a comparative institutional analysis of Brazil and China, with a focus on critical agrifood and rural questions: the formation of a soy–meat complex linking the restructuring of their respective agrifood systems, in respect of both production and consumption; and secondly, discusses the active role of actors and institutions linked to peasants and family farmers in the newly emerging RD dynamics in both countries. The fourth section synthesizes the discussion and provides concluding remarks.

The rise of China and its implications for Brazil

The trajectory of rapid economic growth and development in China over the past 30 years is a most remarkable demonstration of the dynamic nature of capitalism in the age of 'globalization', as well as of the restructuring that China's rise has caused in the dynamics of the world economy. The resurgence of China in the contemporary era coincides with the collapse of 'real socialism' in the USSR and Eastern Europe, which buried the bipolar order of the Cold War, and with the reassertion of the hegemony of the United States, which marked the rise of neoliberalism. If these dramatic late twentieth century events mean that China has embraced globalization, this embrace should not be confused with an unrestricted adherence to neoliberal principles, as Harvey seems to suggest (2005). Nor does China's rise to the status of a great economic power seem to be determined solely by the secular and exogenous macro-structural factors identified by Arrighi (2010), as the systemic cycles of accumulation of wealth and power and the corresponding hegemonic transitions show. It is reasonable to assume that the changing geopolitical context, particularly in relation to institutional configurations and an assertive strategy of state-led development, all played important roles in China's economic trajectory.

When critical voices against neoliberalism began to emerge, Ramo (2004) provocatively proposed the possible replacement of the hegemonic 'Washington Consensus' by an alternative 'Beijing Consensus' (BC). Zhao (2010) argued that whether Ramo's notion of BC and the different uses people make of it are accurate or not, or whether and when China will replace the Western model of modernization and appropriate economic and political hegemony from the US, the 'China model'[1] of state-led economic reforms with limited political reforms has indeed gained ground in many developing countries. Meanwhile, with regard to China's 'going global' internationalization strategy, despite its efforts to 'grow based on the domestic market' at a time when the advanced capitalist economies are in crisis, China will probably continue to pursue two key strategies: to channel industrial exports to the markets of economies with greater potential for expanding consumption; and to redirect its flow of trade and investments to provide sources of supply for the strategic natural resources, energy, raw materials, and foodstuffs it needs (Armony & Strauss, 2012; Cunha, Lélis, & Bichara, 2012). China frames its 'going out' imperative and its 'arriving in' Latin America in terms of the complementarities and mutual benefits that can be gained through 'comparative advantages' and 'division of labour'. Particularly in the Brazilian case, the discourse emphases 'win–win results' and 'mutual cooperation', and a newer emerging rhetoric of global partnership based on equality, which has gained increasing legitimacy from the BRICS institutionalization process since 2009 (Strauss, 2012).

With the US having been displaced as the main centre of gravity of the world economy, the rise of China as a great power brings opportunities and challenges, uncertainties and concerns for Brazil. The data and literature indicate that Brazil is experiencing a set of contradictory impacts associated with the so-called China effect.[2] Direct effects are related to the bilateral flows of investment and trade. Data from the Brazil–China Entrepreneurial Council (CEBC, 2016) show that between 2007 and 2013, US$56.5 billion of Chinese investments entered Brazil, of which US$28.3 billion were confirmed. Between 2014 and 2015 another US$11.4 billion arrived, of which US$9.2 billion were confirmed. Most of these investments were in mergers and acquisitions, primarily in mining, oil and gas, and secondarily, in agribusiness. Investment in agribusiness is particularly controversial, as Chinese companies have purchased vast rural tracts of land and concerns have been expressed around concentration and 'foreignization' of land ownership in a context of 'global land-grabbing'.[3]

However, the main driver of Brazil–China economic relations is trade, which has grown enormously in the decade to 2015 (Figure 1). The most remarkable characteristic has been a notable change in the structure of Brazil's trade with China by value-added and technological content (Table 1). Brazil's exports are mainly primary and resource-based commodities (basically iron ores and soybeans), while its imports are predominantly manufactured goods (mostly machinery, equipment, and electronics).The balance of trade has been favourable for Brazil, with surpluses in primary products and deficits in manufactured goods.

Brazil's economic relations with China have also had a number of indirect effects linked to the terms of trade and competition in third markets. According to Demeulemeester (2012), in the period

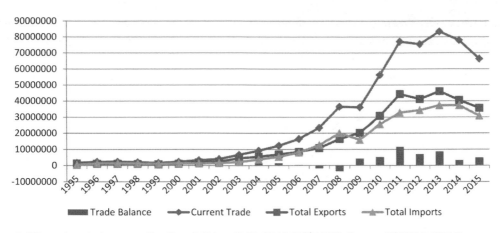

Figure 1. Bilateral trade between Brazil and China, 1995–2015 (US$1000). Source: UNCTAD (2016).

from 2000 to 2010 between 16% and 25% of the total value of the 10 main Brazilian commodities exported are explained as the indirect effect of China's demand on prices, with a significant improvement in the terms of trade. According to Honório and Araújo (2013), the terms of trade index (the ratio between the price indexes of exports and imports), which stayed around 100 between 1999 and 2005, grew to 110 in 2009 and 130 in 2011. With this rise in the prices of its main products, Brazil benefited from exports to China, and to other markets too. However, Brazil is facing fierce competition from Chinese-manufactured goods exported to third markets. After calculating indexes for export concentration, trade complementarity, and constant market share, Demeulemeester (2012) concludes that Brazilian exports have a growing sector concentration in its main destination markets, while Chinese trade is becoming increasingly diversified; and Brazilian trade is becoming less complementary with its main partners, while the Chinese trade is becoming more complementary. Consequently, between 2002 and 2011, Brazil lost market share to China in manufactured exports to the US (–12.87%), the European Union (–5.49%), and Latin America (–7.3%).

Furthermore, these effects have cumulatively had wider consequences related to the risks of 'regressive specialization' in the production, employment, and foreign trade structures of the

Table 1. Brazil–China exports, imports, and trade balances, 2001–2015 (US$1000).

Category of products	Exports (%)			Imports (%)			Balance (US$1000)		
	2001–2005	2006–2010	2011–2015	2001–2005	2006–2010	2011–2015	2001–2005	2006–2010	2011–2015
Primary products	37.7	42.0	49.9	3.3	1.8	2.4	7,540,394.37	34,891,449.94	99,698,285.03
Resource-based products	38.8	47.9	43.4	17.8	10.6	10.4	5,740,108.00	32,686,298.32	72,328,725.17
Low-tech manufactures	9.5	2.6	1.7	13.8	19.6	21.6	86,672.69	–13,887,787.60	–33,814,008.12
Mid-tech manufactures	11.6	5.6	3.4	21.6	26.4	32.1	–587,856.53	–16,804,389.55	–48,327,578.76
High-tech manufactures	2.3	1.8	1.5	43.5	38.7	33.4	–5,649,877.11	–30,203,506.98	–54,486,635.29
Unclassified manufactures	0.1	0.1	0.1	0.1	2.9	0.2	8587.06	–2,278,604.17	–5952.16
Primary products	76.5	89.9	93.4	21.0	12.4	12.7	13,280,502.36	67,577,748.27	172,027,010.20
Manufactures	23.5	10.1	6.6	79.0	87.6	87.3	–6,142,473.89	–63,174,288.30	–136,634,174.33
Total	100	100	100	100	100	100	7,138,028.47	4,403,459.97	35,392,835.87

Source: UNCTAD (2016).

Brazilian economy. Despite the Chinese rhetoric of 'mutual development', there is growing concern in Brazil with regard to two problems: 'reprimarization', given the dependence on exports of agricultural and mineral commodities (now exceeding 50% of the export basket) (Branco, 2013; Delgado, 2012); and 'deindustrialization', apparent in the fact that since its peak, in the 1980s, the GDP share of manufacturing sector has reduced from 33% in 1980 to 16% in 2010, and the total employment share of manufacturing has fallen from 21% in 1994 to 18% in 2010 (Bresser-Pereira, 2010; Gonçalves, 2012). While one cannot blame China for Brazil's economic problems (most of which are for domestic reasons), the deterioration of quality in their bilateral trade seems to reproduce the traditional pattern of core-periphery dynamics long criticized by the UN Economic Commission for Latin America and the Caribbean.

Brazil is positioning itself in a changing international order responding to the rise of China, and it faces the possibility of losing density, diversity, and complexity in its productive and export structures, as well as its capacity to generate industrial employment, innovation, and technological linkages (Cunha et al., 2012). It could be argued that the nature and character of this 'regressive specialization' can be partly understood an outcome of the 'political economy of agribusiness' (Delgado, 2012; Ioris, 2016). This refers to a hegemonic construction anchored on a coalition of power,[4] that articulates the agrarian bourgeoisie, as well as agro-industrial and mineral-extractive capitals, with state credit policy. This coalition sustains a pattern of capital accumulation through the extraction of economic surplus in the primary sector, pursuing the capture of land rent provided by 'natural comparative advantage'. Competitiveness in the exploration of oil and hydropower also relies on a monopoly over natural resources, but productivity gains are mainly the result of advances in technology with strong inter-sectoral linkages, and most of the asset ownership is under public control. The acquisition of land by foreigners (the global land-grabbing) is something different (see Note 3). However, the competitiveness of agricultural and mineral commodity exports is crucially linked to the overexploitation of natural resources and the capture of land rent, capitalized in land prices and appropriated through financial speculation, with land assets held mostly by domestic corporate actors, with many negative consequences on human health and the environment, rural livelihoods, and diversification of farming.

A comparative institutional analysis of agrifood and RD dynamics in Brazil and China and the emergence of a 'countermovement'

This section combines institutional and class analysis to compare the 'agrifood question' and 'rural development dynamics' in Brazil and China. Comparison, according to Byres (1995, p. 572), 'can open analytical perspectives when securely based theoretically, by extending our range of criteria independent of the particular context, and so allowing theory to be more nuanced in what it can reveal'. All the BRICS countries seem to pursue a path of economic development and an international projection of their power which is strongly grounded in leadership in their respective regions, but there are significant disparities between them and unequal capacities to operate as 'regional locomotives', given their distinct economic dynamics (Fiori, 2007; Visentini, 2012). Regarding agrarian issues, these points are treated in papers by Sauer, Balestro, and Schneider (2017) for the case of Brazil in Latin America, and Mills (2017) for the case of China in Southeast Asia.

In the broader sociological literature, the patterns of inequality, social stratification, and class mobility within these countries are also relevant topics, with special attention to the rise of the 'new middle classes' and the recomposition of the working classes. Also important are the entry of working classes into consumer markets, their relations with the dominant classes, and the repercussions of these developments on the social values and political projects that guide development trajectories in the BRICS countries (Peilin, Gorshkov, Scalon, & Sharma, 2013). Meanwhile, the recent convergence of financial, food, energy, and environmental crises has put the nexus between 'food security (and sovereignty)', 'rural development', and 'development in general' back onto the centre stage of theoretical, policy, and political agendas in the world today (Borras, 2009). All these issues have wider implications for BRICS countries, and particularly for Brazil and China.

The agrifood question and the assaults of the 'satanic mills'

Between the biennia 1990–1992 and 2012–2014, the number of chronically undernourished people in the world fell by 39.6%, from 1.014 billion to 805.3 million. Asia and Latin America were the regions that performed best, especially influenced by changes in China and Brazil. In Brazil, between biennia 1990–1992 and 2000–2002, the percentage of hungry people decreased from 14.8% (22.5 million) to 10.7% (19 million) and then fell to less than 5% of the population in 2012–2014, a very low level. In China in the same periods of time, the percentage of hungry people decreased from 23.9% (288.9 million) to 16.1% (211.7 million) and then to 10.6% of the population, a moderately high level (FAO, IFAD, WFP, 2014, pp. 41–43).The paradox that developing countries such as Brazil and China are experiencing is that the old and declining problem of hunger coexists with new and rising problems of malnutrition and obesity. The total percentage of obese and overweight people in Brazil is 63.5% and 27% in China (FAO, IFAD, WFP, 2014). This is linked to the phenomenon of 'nutrition transition' and corresponding changes in eating habits and class-related diets. Large-scale urbanization and improvements in the incomes of 'affluent' portions of the population in developing countries (the 'new' working and middle classes) are accompanied by shifts in lifestyles and changes in diet from large proportions of staple foods and low proportions of protein, based on cereals, vegetables, and fibre, to diets which feature ultra-processed foods and which are low in fibre and high in animal protein, saturated fats, salt, and sugar. There have also been shifts toward lower rates of fertility and mortality, growing incidence rates of chronic and degenerative diseases, and decreasing incidence rates of infections and under-nutrition (Lang & Heasman, 2006).

In sum, the phenomena of increasing financial speculation and land-grabbing, corporate agribusiness control over and appropriation of the value chains in which family farming is inserted, reduced state regulation of agricultural and livestock production, and greater liberalization of foreign trade, could all be seen as expressions of the commodification movement in both Brazil and China. Against this background, the roots of the 'agrifood question' stem from the globalization of developing countries' agrifood systems (McMichael, 2013; Peine, 2009; Wilkinson, 2009).

A most significant issue in understanding the nature and characteristics of the Brazil–China agrifood question is the formation of a 'soy–meat complex' (Weis, 2013), whose dynamics have been a decisive driver in their economic relations over the last 13 years. Indeed, the boom of soybean production and exports in Brazil and the restructuring of the meat and feed industries in China have been two mutually constitutive phenomena which encapsulate in many ways the ongoing changes in their respective agrifood systems, and are emblematic of a wider process of polycentric restructuring of the dynamics of the global food regime. As many authors have shown, the analysis of both the production and consumption sides of this 'complex' can illuminate the ways through which socioeconomic, environmental, health, and cultural concerns intertwine in the two countries (Wilkinson & Wesz, 2013; Schneider, 2014; Oliveira & Schneider, 2016; Schneider and Sharma (2014); Oliveira & Hecht, 2016; Oliveira, 2016; Yan, Chen, & Ku, 2016; Wesz, 2016).

Some data and stylized facts help to illustrate this issue in respect of China. The per capita GDP of China (in purchasing power parity – PPP) increased from US$302 in 1980 to US$1227 in 1992, US $3804 in 2003, US$9012 in 2010, and US$12,879 in 2014 (IMF, 2015). Its urbanization rate, about 18% in 1980, exceeded 50% in 2010 (CSY, 2015). The number of people below the poverty line of US $1.25 a day (PPP) decreased from 63.8% of the Chinese population in 1992 to 28.4% in 2002 and 11.8% in 2009 (World Bank, 2015). Between 1980 and 2010, household expenditure per capita with food (in nominal prices) increased by a factor of 22 in rural areas and 15 in urban areas; and their share of total expenditure decreased from 68% to 41% in rural areas and from 57% to 36% in urban areas during the same years (Garnett & Wilkes, 2014, p. 47).

This socioeconomic transformation of China is a result of institutional changes introduced through the reforms, especially in rural areas (household responsibility system, township, and village enterprises), but also in urban areas (special economic zones and relaxation of the hukou household registration system to permit more migration). These have altered the composition of social classes (especially the new middle class and migrant workers) and fuelled an expansion in the consumer market (particularly food markets). Undoubtedly, the change in eating habits towards a greater consumption of animal protein is the most important factor.[5] China quadrupled its per capita meat consumption from 1980 to 2010, reaching an average of 61 kg per annum, compared to 42 kg in the world as a whole, 71 kg in Brazil, and 120 kg in the US (Oliveira & Schneider, 2016). The country produces and consumes over 50% of the world's pork, 20% of the world's poultry, 10% of the world's beef, and is the fourth largest milk producer (Schneider & Sharma 2014). Pork is a special case: in China, the average per capita consumption is 39 kg per year, compared to 27 kg in the US (Schneider, 2011).

In short, in a context of rapid social change, the adoption of these new diets plays an important role as a practice of distinction and a social marker by defining consumer class identities,

> both symbolically as a representation of status and success, and materially as a high-priced food commodity. For example, an agribusiness executive in Shanghai, discussing development prospects for his firm, is quoted as saying: 'Meat signifies wealth. The more money you have, the more meat you will eat'. In China, meat consumption is increasing across the population, but is most pronounced in urban areas, and among middle-and upper-class consumers. (Schneider, 2014, p. 617)

These changes in consumption have been a key demand driver for the restructuring of the meat industry, and for increasing feed imports in China. While soy-derived foods remain common staples in Chinese cooking and diets (most from still non-GMO [genetic modified organism] domestic soy production), people now consume soy increasingly as cooking oil and in the form of industrial pork or poultry (mostly using GMO-imported beans in animal feeds) (Yan et al., 2016). According to Sharma (2014, p. 14),

> [b]etween 2001 and 2012, China's import dependence doubled from 6.2% to 12.9% with a net deficit in agriculture and food. China's targets for meat production growth are likely to add to that trend in the coming decade due to feed imports.

China is the world's leading importer of soybeans, having bought 60% of the soybeans traded in the international market in 2012. Brazil is its major supplier (45%), ahead of the US (39%), and Argentina (13%).[6] The feed industry is largely controlled by Chinese-owned private companies and 'dragon-headed enterprises' (DHEs), and specialized supply-and-marketing cooperatives control no less than 60% of the soy traded to China, challenging the power of the four major transnational players ADM, Bunge, Cargill, and Dreyfus (collectively known as ABCD) (Sharma, 2014).

There is a similar pattern in the ownership of China's pork industry and operations. Even though it is based on the logic and practices of agribusiness elsewhere, production in the industry is largely domestic. DHEs are said to control – in a certainly overstated estimation – around 70% of pork and poultry production, their operations are supported by large governmental credits and subsidies, and they coordinate a number of stages of the value chain, combining vertical integration and contract farming systems. DHEs have changed the face of pig production in China. In 1985, 'backyard farmers' (typical Chinese peasants) produced at least 95% of the country's pork. But in 2009, the figure was quite different: backyard farmers (1–10 pigs/year) accounted for about 27% of nationwide pig production; 'specialized household farmers' (50–500 pigs/year) (simple commodity producers) accounted for about 51%; and 'large-scale commercial farms' (>500 pigs/year) (entrepreneurial or even capitalist producers) about 22%. While backyard farmers still engage in a kind of agroecological style of farming, both specialized household farms and large-scale commercial farms produce under concentrated animal feeding operation systems controlled by a few large DHEs, which subordinate their practices to the logic of contractual relations (Schneider & Sharma, 2014).

Data and stylized facts about Brazil similarly illustrate fundamental transformations in agrifood systems. Brazil's soybean production grew by 5.5% annually between 2000/2001 and 2013/2014. The harvested grain volume skyrocketed from 13.9 to 87.5 million tons in the period. The national harvested area increased by 4.3% per year, from 14.0 to 30.1 million hectares. Productivity increased by 1.1% per year. Soy currently represents 52.9% of the total grain area and its cultivation is concentrated in the South and Midwest regions, which have the top five producing states in the country. The gross value of production (GVP) of soy in Brazil grew 7.7% per year from 1996 to 2012, increasing the share of GVP in national agricultural GDP from 9.4% to 25.73%. Although there was a 67.3% reduction in poverty between 2003 and 2012, bringing the Engel coefficient for food to less than 16% of average household expenditure and enormously expanding the domestic market, the main target of Brazilian soy is the foreign market. Total Brazilian soybean exports increased at a rate of 8.84% per year between 2000/2001 and 2013/2014 from 15.5 to 44.5 million tons. The soy complex is one of the most significant contributors to the country's trade surplus, growing six-fold between 1997 and 2013 and currently accounting for over 37% of the trade balance of Brazilian agribusiness. A total of 51% of the Brazil's soybean is exported, accounting for 41% of the global market, making Brazil the leading soybean exporter, followed by the US and Argentina (Hirakuri & Lazzarotto, 2014).

In 2011, 36.9% of soy grain entering China came from Brazil and 67.1% of Brazil's soy exports went to China. In 1997, 43% was exported in the form of meal, 43% in oil, and only 14% in beans. From 2000 onwards, exports of meal sharply declined, and 93.4% of the value of the soy complex exports in 2011 was derived from beans. This is partly an outcome of the 'Kandir Law', enacted in 1996, which reduced taxes on the exports of primary products while keeping the regime on processed products the same. This increased Brazil's exports of beans, but reduced profits, as well as the capacity and power of the national crushing industry, adding to the problem of regressive specialization and relative deindustrialization (Wilkinson & Wesz, 2013).

The global food regime within which the Brazil–China soy–meat complex is situated has as its main characteristic the dominance of large transnational corporations, plus a few domestic companies, in every link of agrifood value chains. McMichael (2013) calls this a 'corporate food regime' and Ploeg (2010) uses the term 'food empires' to describe the major agrifood companies and their connections. In-depth documentation of national and transnational corporate control and market power in the soy–meat complex is provided by Wilkinson and Wesz (2013) and Wesz (2014) for Brazil, and Schneider (2014) and Schneider and Sharma (2014) for China.

The point to highlight here is that the literature on the operation of the soy–meat complex in both countries is replete with instances of both negative consequences and the contestations that these generate (Oliveira & Hecht, 2016). In Brazil, the deleterious effects of the soy complex include: the deforestation of the Cerrado and Amazon biomes; degrading labour conditions; monoculture specialization and intensification; high rates of contamination by pesticides and herbicides; and speculation in landed property and inflated land prices, land concentration, and dispossession of family farmers and other rural social groups (indigenous and traditional people) (Delgado, 2012; Oliveira, 2015; Sauer & Leite, 2012). In China, several deleterious effects include: degradation of water and soil; greenhouse gas emissions; biodiversity losses; dietary intake inequalities; and pressures on rural livelihoods leading to dispossession, given the increased commoditization of production and competition for market access (Schneider, 2011, 2014; Yan et al., 2016).

Several emblematic instances that have sparked strong contestations in the politics of consumption against corporate agribusiness in both countries are worthy of comment. Consumers' anxieties about food safety and health concerns related to huge food scandals are certainly the most serious and prominent issue in China, possibly with an even higher profile than environmental problems. The case with the largest impact was certainly that involving infant milk powder tainted with melamine in 2008, which resulted in the death of six children and more than 30,000 patients with kidney stones. So-called avian flu outbreaks are another Chinese scandal which has reverberated worldwide. Since 2003, China has reported cases in every year except 2011. Another food safety scandal was the discovery of illegal additives residues in pork, the so-called lean meat powder (these are drugs banned in China since 2002). The American fast food company KFC recently found suppliers feeding their chickens more than 18 kinds of antibiotics, drugs, and hormones (Schneider & Sharma, 2014).

In Brazil, food safety scandals, industry fraud, and the discovery of contaminants in food are quite common. The abuse of pesticides in agriculture is the most prominent problem, and has triggered intense public debates, civil society mobilization, and a backlash from corporate agribusiness. Although Brazil is the third major agricultural producer in the world (behind the USA and China), it is the leading pesticide consumer: an average of 12 litres of pesticide per ha and 4.5 litres per inhabitant is used in farming. Between 2002 and 2012, the global market for pesticides grew by 93%, while the Brazilian market grew by 190%. There have been recurrent cases of environmental and human contamination. About one-third of vegetables consumed daily are contaminated; the 'poison rain' of pesticides sprayed from aircraft causes a range of health problems; and 10 toxic substances have been found in breast milk in Lucas do Rio Verde town in Mato Grosso do Sul, the major soy-producing state (Carneiro, Rigotto, Augusto, & Búrigo, 2015).

In both countries, there have been significant concerns expressed about current and potential problems related to the cultivation, consumption, and regulation of GMO crops (and traded soy is almost all GMO) (Oliveira & Hecht, 2016). All these negative consequences of the commodification of farming and food, liberalization of trade and globalization of developing countries' agrifood systems led by transnational agribusiness corporations and food empires under the aegis of the current international food regime, have given rise to affected social actors raising the alarm and advocating alternative approaches to food production. This is a strong reason why the new RD dynamics that have emerged in Brazil and China since the 1990s can be interpreted as part of a Polanyian 'countermovement'.

RD dynamics and the emergence of 'countermovements'

For analytical purposes, it is important to distinguish between RD practices, policies, and processes, the basic components of the newly emerging dynamics of RD. While the situations in Brazil and China have their own specificities, various authors have identified some common threads (Hebinck et al., 2015; Ploeg et al., 2012).

RD practices are grassroots-level activities that significantly alter the routines and outcomes of farming. These aim to shift the boundaries of agriculture, create resistance to the control and dependency imposed by food empires, and build autonomy. This includes organic or ecological agriculture, craft processing of high-quality products, building alternative networks, short food-supply chains, ensuring self-provisioning, pluriactivity and access to non-agricultural income sources, reducing production costs, and investing in labour-driven intensification and skill-based technologies. In relation to the wider economic environment in which farmers are located, this implies the inclusion of non-agricultural productive activities into the farm, such as agro-tourism and hospitality services, energy production, mining, aquaculture, nature conservation, and landscape management.

RD policies represent efforts made by the state to respond to the demands and expectations of key social actors, stimulating and strengthening, or restricting and controlling, both the individual practices listed above and general processes of RD. Distinct from traditional agricultural and agrarian policies, based primarily on costing credits and subsidies, the new generation of RD policies focuses mainly on opening and sustaining new circuits of the social and economic reproduction of peasants and family farmers, allowing them to remain viable as food producers, even in the face of adverse market conditions. The main strategic goal of such RD policies is to redefine the role of agriculture in society, but it can embrace and support a wide range of related social goals, such as food security and safety, social and economic equality, and human health and environmental sustainability.

The *modus operandi* of RD processes basically consists of redefining and redirecting the direction of agriculture, food, and rural areas through the regulation of existing markets and the construction of new, 'nested' markets. In practice, it constitutes a way of bypassing the controls exerted by the food empires on the commodity chains and bridging the structural holes they create between producers and consumers. The relationships and interactions within these markets aim to connect previously separated value circuits in order to improve farmers' incomes and livelihoods, offering good quality products at fair prices for consumers, and creating new forms of sociability and reciprocity, trust and shared values for both, even when what they exchange are, as a matter of fact, (food) commodities.

In Brazil, the struggle against poverty and historical inequalities, the promotion of social reproduction and the pursuit of the political emancipation of family farmers and other rural groups constitute the centre of gravity of RD policies and practices, by creating mechanisms to include their participation and valorize their diversity against monocultures and the hegemony of 'agribusiness'.[7] In China, the centre of gravity of RD policies and practices is the maintenance of social stability in an explosive context of economic growth, industrialization, and urbanization, seeking to contain the escalating inequality and promote a 'harmonious society', with rural–urban integration and the inclusion of peasant migrant workers as citizens. In both countries, RD dynamics become more effective when they can synergistically combine government policy interventions with the actors' initiatives and practices, and support of farmers' organizations and social movements, in order to adjust the institutions to the social values in a given balance of forces (Ploeg et al., 2010, 2012).

For a long time, RD in Brazil was associated with a set of compensatory interventions by the state and international agencies in poor rural areas that had failed to integrate the technology patterns of the 'agricultural modernization' process, along the lines of the 'Green Revolution' of the 1960s.

However, by the 1990s, under a contradictory combination of re-democratization and neoliberalism, many social actors repressed during the dictatorship returned to the scene and new civil society actors began to come to light. It is in this context that RD emerged as an important theme. After that, family farming came to be in the forefront of public debates about agriculture and food production and began to assume greater legitimacy as a social category against the background of struggles for political recognition driven by a range of actors: farmers' unions and rural social movements and organizations like CONTAG (Confederation of Agricultural Workers), MST (Landless Rural Workers' Movement), and FETRAF (Federation of Workers in Family Farming).

This spurred interest among scholars, researchers, mediators, and advisors and led to a surge of academic studies between 1990 and 1996 that produced interpretative displacements and influenced policy-making, leading to institutional recognition from the state and the development of new RD policies. The struggles of organized social movements and their pressure on the state led to the establishment of the most important RD policy instrument, PRONAF (National Program for Strengthening Family Farming), in 1996, focusing on credit for working capital and investment. Negotiations between social actors and government have increased the resources available for the Family Farming Harvest Plan every year. In 1998, the Ministry of Agrarian Development was established and a Family Farming Law was passed in 2006. CONDRAF (National Council for Sustainable Rural Development) and CONSEA (National Council for Food and Nutritional Security), both established in 2003, are also important spaces for the political representation of RD actors.

Two other RD policies deserve special note. The Food Acquisition Program was created in 2004 to respond to farmers' problems with commercialization, price instability and market access, and to meet the demand for food from public institutions (hospitals, prisons, army, etc.). The PNAE (National School Feeding Program) was reorganized and obligations placed on municipalities to purchase at least 30% of school feeding products from family farmers' new institutional markets.

The evolution of RD policies in Brazil involves three co-evolving generations of such policies. The period from 1993 to 1998 was characterized by the emergence of an agrarian policy involving land reform, rural settlements, and differentiated rural credit policies. The period from 1998 to 2004 was characterized by widely implemented compensatory and distributive policies focusing on rural social assistance and protection. From 2005 up to the present day, policies began to incorporate a strategy of building new markets to ensure food and nutrition security, with enhanced channels of connection between consumers and farmers, and an attempt to promote environmental sustainability through specific norms, regulations, and incentives (Schneider, 2010; Schneider & Cassol, 2013; Schneider, Shiki, & Belik, 2010).

In China, the institutional reforms of the late 1970s and early 1980s that targeted agriculture and the countryside were welcomed and increased production, productivity, and incomes, thus reducing rural poverty. However, in the passage from the 1980s to the 1990s, a sense of dissatisfaction with stagnating and deteriorating living conditions in the rural areas and a massive increase in rural–urban migration became apparent. This crisis became known as 'sannong wenti', the three agrarian problems: agriculture (*nongye*), countryside (*nongcun*), and peasants (*nongmin*). This concept is a hallmark of the new RD dynamics in China, and emerged to explain the limitations of the growth of agricultural production, the limitations of improving the living conditions in the countryside, and the difficulties of enhancing the income and well-being of the farmers (Ye, Rao, & Wu, 2010).

Rural change processes in China involves leadership and negotiation between different social actors: formal representatives of the party-state at various levels (village, township, city, province, central); social activists such as academics and intellectuals linked to universities or NGOs; local residents, through grassroots organizations (farmers' cooperatives, religious, and cultural associations);

or rural elites (business and local party cadres) (Thøgersen, 2009). During the 1990s, a so-called new left emerged in Chinese politics, proving to be a key actor within rural debates and adopting strong positions against the privatization of land and in favour of the protection of peasant livelihoods from market forces. Its most prominent manifestation is the New Rural Reconstruction Movement (NRRM), an attempt to bring together rural experiments in democratic cooperatives and social organization modelled in part on the Rural Reconstruction Movement of the 1930s (Day, 2013).

The state also took steps to address the *sannong wenti*. The 'tax-for-fee' reform in 2002 foreshadowed the abolition of agricultural taxes that was completed in 2006. In the same year, a new programme called 'New Socialist Countryside' (NSC) was launched to promote RD, reduce income inequality, and prevent social unrest, by redistributing resources through fiscal transfers from the central to county level, and then to villages and township projects at the local level, focused especially on infrastructure and social services. Despite the shared aim of addressing rural problems, the thrust of the state's programme is different from that advanced by left-wing intellectuals, insofar as it calls for further urbanization, consumption, and market-driven growth. While the NSC has meant different things in different areas, one important feature is the encouragement of in-place urbanization and commercialization, making use of specialized supply-and-marketing cooperatives and DHEs which recognize, to some extent, that migration to large cities is not an immediately viable path for the majority of China's peasants (Ahlers & Schubert, 2009; Day, 2013; Yeh, O'Brien, & Ye, 2013).

In China, RD occurs in a pragmatic and incremental manner. The government, at different levels (central and local), has the capacity to recognize and endorse promising practices and initiatives by social actors at the grassroots level. The state then supports these experiences, including them within its broader strategy, mobilizing the necessary actors to implement it and institutionalizing a policy framework which is quite flexible and adaptable to practical contingencies (Ye et al., 2010). In Brazil, on the other hand, the structural holes created during the period of agricultural modernization and the dismantling of state institutions and policies in the neoliberal period have been filled by the inclusion of civil society in the formulation and operation of actions, programmes and policies. To the extent that the state has become more permeable to social demands, the scope of action of the rural social movements and organizations also seems to have shifted over time, from a vindictive and anti-establishment stance of the 1980s to a purposeful and proactive stance in the 1990s and the co-management of public policies by the mid-2000s (Schneider et al., 2010). Be that as it may, despite their differences and singularities, the *modus operandi* of RD processes in both countries take place through regulating old markets and constructing new, nested markets to provide food security for urban residents and support circuits of social reproduction for farmers' livelihoods.

There is no space here for a more detailed empirical analysis of RD dynamics in each country. However, two theoretical clarifications are unavoidable to make sense of the 'double movement' thesis advocated in this paper. The first is on the heterogeneity of family farming. The debate goes back to the endless controversy among Marxists, followers of Lenin's thesis, and Substantivists, followers of Chayanov's thesis and referred to as 'populists' by their detractors. This is not the place to take up this debate, reproduced in China and Brazil, but simply to state a position. For the authors of this paper, both approaches have key insights, though neither is fully correct. Marxists are right to stress that one cannot see family farming as an undifferentiated whole with homogeneous interests, but wrong to point out the trends of capital accumulation (from above) and class differentiation (from below) leading to a inexorable 'disintegration of the peasantry' without taking into account the agency of rural actors' livelihoods reproduction strategies and the complex and differentiated character of the commodification of agriculture. And Chayanovians are wrong to overlook the internal differentiations of family farming (or 'the peasantry'), but right to emphasize the difference between the family form of organization of labour and production in agriculture and the capitalist form based on wage labour.

Utilizing insights from both schools of thought, the authors of this paper subscribe to Ploeg's (2009) notion of 'farming styles' to distinguish at least three segments of family farming with different degrees of commodification of the means of production and livelihoods and insertion in (conventional or nested) product markets: a typical peasant style of domestic subsistence producers; a commercial, integrated style of petty commodity producers (PCP); and an entrepreneurial style of simple commodity producers, to use the terms of the 1980s' Marxist debate. The main differences amongst them is in relation to the intensity of their 'peasant condition' and their 'degree of commodification', basically perceived through the proportion between the *quantum* of commodities mobilized in various markets and the use values reproduced internally within the production unit (a 'self-controlled resource base'). Thus, in line with Wolf's (1984) contention that the 'middle peasantry' was the most revolutionary actor in the 'peasant wars' of the twentieth century, strengthening the autonomy of a 'commercial style' (PCP) of family farming and constructing new, nested markets with a better distribution of income and wealth are crucial tasks for RD dynamics in the twenty-first century (even if it cannot be considered a revolutionary task).

The second clarification relates to the politics of food consumption and RD dynamics. The relationship between producers and consumers in the construction of nested markets and alternative food networks is a key topic in both China (Si, Schumilas, & Scott, 2015; Wang, Si, Ng, & Scott, 2015) and Brazil (Niederle, 2014; Radomsky, 2013). Anxieties related to food safety scandals and contaminations of food due to the abuse of pesticides are the main drivers of changes in eating habits and consumption behaviour in food markets. In China, these generated state responses in form of the institutionalization of certification schemes for ecological products (green food, organic food, and hazard-free food) and, in Brazil, to the rise of social movements promoting agroecological farming, farmers' markets, networks of producers and consumers (e.g. Ecovida Agroecology Network, *Rede Ecovida de Agroecologia*), and collective food-buying groups. The sceptical attitude of Chinese consumers towards state regulation led to the rise of self-organized, community-supported agriculture, farmers' markets, and buying groups without the need of certification (mostly supported by the NRRM). In Brazil, the struggles of social movements led to the institutionalization of agroecology through public recognition of diversified certification schemes (third party, participatory, and direct sales without seal). In both cases, through different paths, the result was the coexistence of official schemes based on institutional trust (sometimes challenged) and alternative schemes based on consumer–producer reciprocity and interpersonal and organizational trust.

Of course, as in developed countries, as well as in Brazil and China, there are debates about the 'conventionalization' (Guthman, 2004) of ecological farming and alternative food networks, as well as the motives of consumers, who are more motivated by health concerns than environmental and social justice. One can argue that the values and practices underlying how society can offer solutions to the agrifood question are in dispute, not in a manicheistic way, but as a struggle between hegemonic and counter-hegemonic forces.

Conclusion

In the first part of this article, the authors discussed the impacts of the rise of China on Brazil's development prospects. While the surge of FDI between these countries since the 2008 global financial crisis constitutes a trend, the main driver of their bilateral relations has been trade. The data presented and literature cited here indicate that Brazilian exports are increasingly concentrated in primary and resource-intensive commodities, and its imports from China are basically manufactured goods with increasing technological sophistication. So, as a major producer and exporter of primary (agricultural and mining) commodities and a diversified producer of manufactured goods, Brazil has experienced both positive and negative impacts from the 'China effect': increased demand and higher prices on the one hand, and competitive pressure on its domestic and export markets on the other hand, which may carry the risk of regressive specialization and deindustrialization. Having outlined this context, the second part of the paper provided a comparative analysis of the agrifood question and RD dynamics in Brazil and China.

Table 2 summarizes the main similarities and differences in both countries.

Through Burawoy's (2003) 'Polanyi-Gramsci framework', we view the agrifood question and RD dynamics in China and Brazil as central issues of the contemporary 'double movement'. We focus on the ways in which key social actors engage with markets and politics to attempt to turn the balance of forces in their favour and push a counter-hegemonic project against hegemonic vested interests. On the one hand, the commodification movement could be named 'Gramsci's moment'. This is the moment when the economic interests of the food empires and agribusiness capital encounter the interests of the state, being translated as broader societal interests and becoming a constitutive part of the coalition of power that forms a hegemonic bloc. Central to the relationships between Brazil and China is the convergence of hegemonic interests from both sides around trade and investment flows in the formation of a soy–meat complex, with narratives for the justification of vested interests in both sides.

In the Brazilian case, the interest of agribusiness is to pursue a private and state strategy of appropriation of land rent deriving from the comparative advantage of natural resources as the frontline of capital accumulation for the whole economy, articulating large rural property and land markets,

Table 2. Comparison of agrifood and RD issues in China and Brazil – the 'double movement'.

	China	Brazil
The agrifood question		
Similarities	Commodification and globalization of agrifood systems; nutrition transition; consolidation, oligopolization and Internationalization of food empires; formation of a soy–meat complex driving a polycentric restructuring of food regime dynamics; contradictions and negative social, health and environmental consequences	
Differences	Diversification of production and export structures of economy; restructuring of meat and feed industries; China's agrifood system as an importing pole; reduction of hunger to a moderately high level and increase of obesity and overweight to medium level; consumers' anxieties about food safety and health concerns related to huge food scandals	Regressive specialization of production and export structures of economy; boom of soybean production and exports; Brazil's agrifood system as exporting pole; reduction of hunger to a very low level and increase in obesity and overweight to a high level; consumers' concern about environment and health concerns related to the abusive use of pesticides (agrotoxics) in agriculture
RD dynamics		
Similarities	Emergence of actors' practices, state's policies and processes linked to livelihood strategies of peasants and family farmers, alignment with consumers and construction of new 'nested markets' for food security	
Differences	Maintenance of social stability in a context of economic growth, industrialization, and urbanization, seeking to contain inequality and promote a 'harmonious society', with rural–urban integration and inclusion of peasant migrant workers as citizens; government ability to recognize and endorse initiatives at the grassroots level	Fighting poverty and inequality, promotion of social reproduction and pursuit of political emancipation of family farmers and other rural groups, seeking to promote social participation and valorization of their diversity against the hegemony of agribusiness; institutional voids been filled by the inclusion of civil society in the formulation and operation of actions, programmes and policies

agriculture and the downstream and upstream sides of the value chains, and access to finance capital under state credit policy. The claim of universality of such interests is based on a discourse that praises the capacity of agribusiness to generate the trade surpluses necessary to bridge the gap in the balance of payments caused by deficits in the (foreign capital) services account, supposedly crucial to prevent imbalances in the current transactions account. But despite the strength of the state and private ideological apparatuses, this narrative has been challenged by a reality in which the trend of high commodity prices in the international market cannot be taken for granted and Brazil is facing the increasing risk of regressive specialization (Delgado, 2012).

In the Chinese case, a strategy that includes the industrialization of agriculture, large-scale (domestic and overseas) land investments, and land transfers from peasants to agribusiness ventures, all supported by state credit and subsidies, represents hegemonic interests in the state and private sector. These are then presented as the interests of the whole society, built upon two combined narratives: a 'crisis narrative' – 'China is feeding 21% of the world's population on 9% of its arable land', and a 'victory narrative' – 'We will feed ourselves!'. However, the ideological legitimacy of this discourse could be questioned based on the fact that the concept of 'food security' (*shipin fangyu anquan*) is unusual in China, as the concept in current use is 'grain security' (*lingshi anquan*), or grain self-sufficiency, which means adhering to the policy goal of a 95% baseline for grain from domestic production. The massive amounts of soybeans imported to feed pigs to sustain the 'meatification' of diets is not accounted for in the balance sheets of grain security, and there are fears the same could happen with maize, which is also subject to liberalization and exclusion from the list of 'strategic crops' (Schneider, 2011).

Thus, the cases of both Brazil and China present evidence of the fragility of those agrifood hegemonic discourses coalesced around the interests of the soy–meat complex. A protective counter-movement, or 'Polanyi's moment', is the moment when the actors affected by the first movement (especially farmers and consumers) become aware of the negative consequences of the commodification of food production, distribution, and exchange at the expense of livelihoods, health, and the environment. In this way, new commitments can be created, and RD, food security, and ecological sustainability can be asserted as constitutive parts of a counter-hegemonic project able to represent the broader societal interests of different classes, groups, and segments of the population, in order to respond and react to the new challenges and lead society on a path towards the construction of a new historical bloc.

Both in Brazil and China, the manner in which these values and commitments evolve in terms of RD practices and policies depends on interactions between certain categories of actors and institutions: the organized peasants and farmers; the organic intellectuals; some policy-makers; and, increasingly, consumers. The general interest of such an alliance is to meet the expectations of those affected by the commoditization of food, by ensuring access to food and food quality for urban and rural citizens as consumers, and to create circuits of social reproduction and improved livelihoods for the farmers as producers. The mechanisms by which these outcomes have been achieved in both countries involve two combined processes: the creation of public regulations and controls over conventional markets; and the active social construction of new, nested markets. Such processes can open new pathways through the reconstitution of existing patterns of production, but principally of circulation, distribution, and the use of resources in ways that dominant interests find it difficult to capture.

In sum, the meaning of the twenty-first century's 'double movement' might be interpreted as follows. The movement led by agribusiness capital and food empires for the commodification, liberalization, and globalization of the agrifood systems of Brazil and China under the current food regime is understood as 'Gramsci's moment'. Gramsci correctly saw the realm of production as the site of organization of the interests of the capitalist class and capital accumulation as the source of power and the material basis for their hegemony. But as for Gramsci, along with Marx, markets are just the space of realization of capital, and commodity fetishism is a veil that obscures the productive core of capitalism and uncovers its contradictions. Both are limited in thinking that only the realm of production can provide the basis for counter-hegemony, and thus discard the possibilities of a producer–consumer political alliance and the agency of actors in the realm of circulation, leaving little space for countermovements.

The protective countermovement, driven by the actors and institutions affected by the commodification process, can be understood as 'Polanyi's moment', precisely because Polanyi was able to conceive the realm of circulation and the relationships between producers and consumers that take place in the markets as potential grounds for counter-hegemony. As Gramsci may warn, of course such an alliance on its own is insufficient, and unlikely to radically overturn capitalist domination as a whole in a 'war of movement'. However, it could wield sufficient power to politicize the contradictions ingrained in these transforming agrifood systems and to shape the direction of RD dynamics, in a 'war of position'. Thus, although the politics of both production and labour relations are crucial sites of struggle in Brazil and China, the construction of new, nested markets, by linking the provision of material needs to shared social values held by both producers and consumers, means that the *locus* of the relationships between them is principally in the realm of circulation. It is here that the transformational potential of an alliance that defends the resilience and autonomy of farmers and the rights of consumers against the 'assaults' of agribusiness and food empires is situated.

Notes

1. Another vision of the 'China model' and the 'lessons' that other developing countries, especially those of Latin America, can learn from the Chinese experience can be found in the works of the former president of the World Bank, Justin Yifu Lin and his collaborators (Lin & Treichel, 2012), and several others.
2. There is a significant literature in Brazil and elsewhere about the effects of China's rise on the Brazilian economy and other Latin American economies. For our purpose, it is sufficient to cite some works that document and analyze the stylized facts briefly outlined in this paper: Sauer et al., 2017; Jenkins, 2015; Jenkins & Barbosa, 2012; Curado, 2015; Ray & Gallagher, 2015; Cunha et al., 2012; Medeiros, 2011; Cano, 2012; Armony & Strauss, 2012.
3. The main controversies involve: the 'developmental outsourcing' purpose of Chinese overseas land-based investments, in which the state plays a key role (Hofman & Ho, 2012); the extent, character, origins, and directions of land-grabbing in Latin America (Borras, Franco, Goméz, Kay, & Spoor, 2012); the impacts of FDI land purchases on land prices, land market dynamics, and land concentration in Brazil; the motivations and interests of the investors; the expansion of the agricultural frontier driven by soy and sugarcane, forestry plantation, cattle ranching and mineral extraction; and the legal measures and regulatory controls taken by the Brazilian state to limit foreigners' access to land (Sauer & Leite, 2012; Wilkinson, Reydon, & Di Sabbato, 2013).
4. The origins of this coalition go back the 1999 currency crisis, when Cardoso saw the activation of primary exports as a strategy of adjustment to the neoliberal order, able to generate trade surpluses to support the balance of payments and control inflation. In the politico-institutional sphere, this position is represented by the so-called *Bancada Ruralista* (Ruralist Block), officially the Agriculture Parliamentary Front, a set of parliamentary representatives from across the party political system ranging from the far right-wing to the center left-wing, united by their defense of agribusiness. Although some of these representatives supported the progressive PT (Workers' Party) government (in itself a contradictory position), most have historically been identified as conservative or even reactionary.
5. An oversimplified representation of this phenomenon is that the standard Chinese food-consumption pattern of 8:1:1, or eight parts grain, one part meat/ poultry/fish, and one part vegetables/fruit, has

been changing rapidly towards a 4:3:3 pattern of four parts grain, three parts meat/fish (and eggs and milk), and three parts vegetables/fruit (Huang, 2011, p. 4).

6. Wilkinson and Wesz (2013) stress that the Chinese interest in acquiring land in Brazil (sometimes with investments in crushing plants and port terminals) is closely linked to its 'need for feed' and its strategy to exert direct control over the soy commodity complex. Schneider (2014) describes such land deals and agribusiness investments as 'meat grabs' rather than using 'food security land grabs', the more common term in the literature. However, as Oliveira (2015) notes, while Chinese–Brazilian value flows around the soy complex are displacing North Atlantic agribusiness power in a South/ East direction, the 'old hubs' of capital (US, EU, Japan) still play very important roles.

7. Since 2013 and the 'Journeys of June', and even more strongly after the impeachment (or parliamentary coup) of 2016 that ousted President Dilma Rousseff and put Michel Temer in her place, Brazilian political and policy landscapes are going through a period of regressive changes producing great uncertainty. But this period is not covered in our analysis.

Disclosure statement

No potential conflict of interest was reported by the authors.

Funding

The work undertaken by Fabiano Escher in researching and writing this article would not have been possible without a doctoral (2012–2016) scholarship provided by the Brazilian Higher Learning Agency Capes (Coordenação de Aperfeiçoamento de Pessoal de Nível Superior), and also a BICAS small grant for which he is grateful.

References

Ahlers, A. L., & Schubert, G. (2009). Building a socialist countryside: Only a political slogan? *Journal of Chinese Current Affairs*, *39*(4), 35–62.

Armony, A. C., & Strauss, J. C. (2012). From going out (*zou chuqu*) to arriving in (*desembarco*): Constructing a new field of inquiry in China–Latin America interactions. *The China Quarterly*, *209*, 1–17.

Arrighi, G. (2010). *Adam Smith em Pequim: Origens e fundamentos do século XXI* [Adam Smith in Beijing: Origins and foundations of the 21st century]. São Paulo: Boitempo.

Arrighi, G., & Silver, B. (2003). Polanyi's 'double movement': The belle époques of British and US hegemony compared. *Politics and Society*, *31*, 325–355.

Borras, S. M. (2009). Agrarian change and peasant studies: Changes, continuities and challenges: An introduction. *Journal of Peasant Studies, 36*(1), 5–31.

Borras, S. M., Franco, J. C., Goméz, S., Kay, C., & Spoor, M. (2012). Land grabbing in Latin America and the Caribbean. *Journal of Peasant Studies, 39*(3–4), 845–872.

Branco, R. S. (2013). Raul Prebisch e o desenvolvimento econômico brasileiro recente liderado por *commodities* [Raul Prebisch and recent Brazilian commodity-led economic development]. *Sociais e Humanas, 26*(1), 197–207.

Bresser-Pereira, L. C. (Ed.). (2010). *Doença holandesa e indústria* [Dutch disease and industry]. Rio de Janeiro: Editora FGV.

Burawoy, M. (2003). For a sociological Marxism: The complementary convergence of Antonio Gramsci and Karl Polanyi. *Politics and Society, 31*, 193–261.

Byres, T. J. (1995). Political economy, the agrarian question and the comparative method. *Journal of Peasant Studies, 22*(4), 561–580.

Cano, W. (2012). A desindustrialização no Brasil. [The deindustrialization in Brazil]. *Economia e Sociedade, 21* (Special issue), 831–851.

Carneiro, F. F., Rigotto, R. M., Augusto, L. G. S., & Búrigo, A. C. (2015). Dossiê Abrasco: *Um alerta sobre os impactos dos agrotóxicos na saúde* [Alert on the health impacts of agrochemicals]. São Paulo: Expressão Popular.

CEBC – Conselho Empresarial Brasil-China. (2016). *Investimentos Chineses no Brasil* [Chinese investments in Brazil] *(2014–2015)*. Rio de Janeiro: CEBC.

CSY (China Statistical Yearbook). (2015). Beijing: China Statistics Press.

Cunha, A. M., Lélis, M. T. C., & Bichara, J. S. (2012). O Brasil no espelho da China: Tendências para o período pós-crise financeira global [Brazil in the mirror of China: Tendencies in the post-global financial crisis period]. *Revista de Economia Contemporânea, 16*(2), 208–236.

Curado, M. (2015). China rising: Threats and opportunities for Brazil. *Latin American Perspectives, 205, 42*(6), 88–104.

Day, A. (2013). A century of rural self-governance reforms: Reimagining rural Chinese society in the post-taxation era. *Journal of Peasant Studies, 40*(6), 929–954.

Delgado, G. C. (2012). *Do capital financeiro na agricultura à economia política do agronegócio: Mudanças cíclicas em meio século (1965–2012)* [From financial capital in agriculture to the political economy of agribusiness: Cyclical changes over half a century (1965–2012)]. Porto Alegre: Editora da UFRGS.

Demeulemeester, J. M. (2012). *Ascensão chinesa: Uma análise de seus impactos sobre o Brasil* [China's rise: An analysis of its impacts on Brazil]. Porto Alegre: Universidade Federal do Rio Grande do Sul (Monografia Graduação em Relações Internacionais).

Evans, P. (2008). Is an alternative globalization possible? *Politics and Society, 36*, 271–305.

FAO (Food and Agriculture Organization of the United Nations), IFAD (International Fund for Agricultural Development), & WFP (World Food Programme). (2014). *The state of food insecurity in the world 2014: Strengthening the enabling environment for food security and nutrition.* Rome: FAO.

Fiori, J. L. (2007). A nova geopolítica das nações e o lugar da Rússia, China, Índia, Brasil e África do Sul [The new geopolitics of nations and the place of Russia, China, India, Brazil and South Africa]. *Oikos – Revista de Economia Heterodoxa, 8*(6), 77–106.

Friedmann. (2009). Discussion: Moving food regimes forward: Reflections on symposium essays. *Agricultureand Human Values, 26*(4), 335–344.

Garnett, T., & Wilkes, A. (2014). *Appetite for change: Social, economic and environmental transformations in China's food system.* Oxford: Food Climate Research Network.

Gonçalves, R. (2012). Governo Lula e o nacional-desenvolvimentismo às avessas [The Lula government and reverse-national developmentalism]. *Revista de Economia Política, 31*, 5–30.

Gramsci, A. (2002). *Cadernos do cárcere Vol. 5* [Prison diaries Vol. 5]. Rio de Janeiro: Civilização Brasileira.

Guthman, J. (2004). The trouble with 'organic lite' in California: A rejoinder to the 'conventionalisation' debate. *Sociologia Ruralis, 44*, 301–316.

Harvey, D. (2005). *A brief history of neoliberalism.* New York, NY: Oxford University Press.

Hebinck, P., Ploeg, J. D., & Schneider, S. (2015). *Rural development and the construction of new markets.* New York, NY: Routledge.

Hirakuri, M. H., & Lazzarotto, J. J. (2014). *O agronegócio da soja nos contextos mundial e brasileiro* [Soybean agribusiness in the global and Brazilian contexts]. Londrina: Embrapa Soja.

Hofman, I., & Ho, P. (2012). China's 'developmental outsourcing': A critical examination of Chinese global 'land grabs' discourse. *Journal of Peasant Studies, 39*(1), 1–48.

Honório, M., & Araújo, M. P. (2013). Corrente de comércio do Brasil: Rumos e desafios [Trade chains in Brazil: Directions and challenges]. *Revista Ciências Sociais em Perspectiva* (Online), *13*(25). doi:10.5935/rcsp.v13i25.9635.

Huang, P. C. C. (2011). China's new age small farms and their vertical integration: Agribusiness or co-ops? *Modern China, 37*(2), 107–134.

IMF – International Monetary Fund. (2015). *World economic outlook database.* Washington, DC. Retrieved October 7, 2015, from http://www.imf.org/external/pubs/ft/weo/2015/01/weodata/weoselgr.aspx

Ioris, A. A. R. (2016). The politico-ecological economy of neoliberal agribusiness: Displacement, financialisation and mystification. *Area, 48*(1), 84–91.

Jenkins, R. (2015). Is Chinese competition causing deindustrialization in Brazil? *Latin American Perspectives, 205, 42*(6), 42–63.

Jenkins, R., & Barbosa, A. F. (2012). Fear for manufacturing? China and the future of industry in Brazil and Latin America. *The China Quarterly, 209*, 59–81.

Lang, T., & Heasman, M. (2006). *Food wars: The battle for minds, mouths and markets.* London: Earthscan.

Levien, M., & Paret, M. (2013). A second double movement? Polanyi and shifting global opinions on neoliberalism. *International Sociology, 27*(6), 724–744.

Lin, J. Y., & Treichel, V. (2012). *Learning from China's rise to escape the middle-income trap: A new structural economics approach to Latin America.* Washington, DC: World Bank (Policy Research Working Paper; no. 6165.).

McMichael, P. (2013). *Food regimes and agrarian questions.* Winnipeg: Fernwood Publishing.

Medeiros, C. A. (2011). A dinâmica da integração produtiva asiática e os desafios à integração produtiva no Mercosul. [The dynamics of Asian productive integration and the challenges of Mercosul productive integration]. *Análise Econômica, 29*(55), 7–32.

Mills, E. N. (2017). Southeast Asia as a hotspot for Chinese investment: A key piece of the puzzle in understanding global land deal trends. *Globalizations.*

Niederle, P. A. (2014). Os agricultores ecologistas nos mercados para alimentos orgânicos: Contramovimentos e novos circuitos de comércio. *Sustentabilidade em Debate, 5*(3), 79–97.

Oliveira, G. (2016). The geopolitics of Brazilian soybeans. *Journal of Peasant Studies, 43*(2), 348–372.

Oliveira, G., & Hecht, S. (2016). Sacred groves, sacrifice zones and soy production: Globalization, intensification and neo-nature in South America. *Journal of Peasant Studies, 43*(2), 251–285.

Oliveira, G., & Schneider, M. (2016). The politics of flexing soybeans: China, Brazil and global agroindustrial restructuring. *Journal of Peasant Studies, 43*(1), 167–194.

Peilin, L. I., Gorshkov, M. K., Scalon, C., & Sharma, K. L. (Ed.) (2013). *Handbook on social stratification in the BRIC countries: Change and perspective.* New York: World Scientific.

Peine, E. (2009). *The private state of agribusiness: Brazilian soy in the frontier of a new food regime* (Unpublished doctoral dissertation). Ithaca, NY: Cornell University.

Ploeg, J. D. (2008). *Camponeses e impérios alimentares.* Lutas por autonomia na era da globalização [Peasants and food empires: Struggles for autonomy in the age of globalization] Porto Alegre: Editora da UFRGS. (Série Estudos Rurais).

Ploeg, J. D. (2009). O modo de produção camponês revisitado [Peasant mode of production revisited]. In S. Schneider (Org.), *A diversidade da agricultura familiar* [The diversity of family farming] (2nd ed, pp. 13–56). Porto Alegre: Editora da UFRGS.

Ploeg, J. D. (2010). The food crisis, industrialized farming and the imperial regime. *Journal of Agrarian Change, 10*(1), 98–106.

Ploeg, J. D. (2011). Trajetórias de desenvolvimento rural: Uma pesquisa comparativa internacional [Rural development trajectories: Comparative international research]. *Sociologias, 13*(27), 114–140.

Ploeg, J. D., Ye, J., & Schneider, S. (2010). Rural development reconsidered: Building on comparative perspectives from China, Brazil and the European Union. *Rivista di Economia Agraria, 65*(2), 163–190.

Ploeg, J. D., Ye, J., & Schneider, S. (2012). Rural development through the construction of new, nested, markets: Comparative perspectives from China, Brazil and European Union. *Journal of Peasant Studies*, *39*(1), 133–173.

Polanyi, K. (1977). *The livelihood of man*. London: Academic Press.

Polanyi, K. (2000). *A grande transformação*. Rio de Janeiro: Elsevier.

Radomsky, G. (2013). Certificações, sistemas participativos de garantia e agricultura ecológica: aspectos da relação entre agricultores e consumidores [Certifications, participatory systems of guarantee and ecological farming: features of the relationships between farmers and consumers]. NIEDERLE, P.A.; ALMEIDA, L.; VEZZANI, F.M. (Org.). *Agroecologia*: práticas, mercados e políticas para uma nova agricultura. Curitiba: Kairós.

Ramo, J. C. (2004). *The Beijing consensus*. Washington, DC: The Foreign Policy Centre.

Ray, R., & Gallagher, K. (2015). *China-Latin America economic bulletin 2015 edition*. Global Economic Governance Initiative (Working Paper 2015, No. 9).

Sauer, S., & Leite, S. (2012). Agrarian structure, foreign investment in land, and land prices in Brazil. *Journal of Peasant Studies*, *39*(3–4), 873–898.

Sauer, S., Balestro, M., & Schneider, S. (2017). The ambivalent and shaky stance of Brazil as a regional power in Latin America. *Globalizations*.

Schneider, S. (2010). Situando o desenvolvimento rural no Brasil [Locating rural development in Brazil]. *Revista de Economia Política*, *30*(3), 511–531.

Schneider, M. (2011). *Feeding China's pigs: Implications for the environment: China's smallholder farmers and food security*. Washington, DC: Institute for Agriculture and Trade Policy.

Schneider, M. (2014). Developing the meat grab. *Journal of Peasant Studies*, *41*(4), 613–633.

Schneider, S., & Cassol, A. (2013). *A agricultura familiar no Brasil* [Family farming in Brazil]. Porto Alegre: FIDA (International Fund for Agricultural Development). (Relatório Pobreza y Desigualdad. Contracto de consultoria de investigación código 2013-05FLI.).

Schneider, S., & Escher, F. (2011). A contribuição de Karl Polanyi para sociologia do desenvolvimento rural [Karl Polanyi's contribution to the sociology of rural development]. *Sociologias*, *13*(27), 180–219.

Schneider, S., Shiki, S., & Belik, W. (2010). Rural development in Brazil: Overcoming inequalities and building new markets. *Rivista di Economia Agraria*, *65*(2), 225–260.

Schneider, M., & Sharma, S. (2014). *China's pork miracle? Agribusiness and development in China's pork industry*. Washington, DC: Institute for Agriculture and Trade Policy (Global Meat Complex: The China Series).

Sharma, S. (2014). *The need for feed: China's demand for industrialized meat and its impacts*. Washington, DC: Institute for Agriculture and Trade Policy (Global Meat Complex: The China Series).

Si, Z., Schumilas, T., & Scott, S. (2015). Characterizing alternative food networks in China. *Agriculture and Human Values*, Amsterdam, *32*, 299–313.

Somers, M. R., & Block, F. (2005). From poverty to perversity: Ideas, markets and institutions over 200 years of welfare debate. *American Sociological Review*, *70*, 260–287.

Strauss, J. C. (2012). Framing and claiming: Contemporary globalization and 'going out' in China's rhetoric towards Latin America. *The China Quarterly*, *209*, 134–156.

Thøgersen, S. (2009). Revisiting a dramatic triangle: The state, villagers, and social activists in Chinese rural reconstruction projects. *Journal of Current Chinese Affairs*, *38*(4), 9–33.

United Nations Conference on Trade and Development – UNCTAD. (2016). *Statistics*. Washington, DC. Retrieved October 11, 2016, from http://unctadstat.unctad.org/wds/ReportFolders/reportFolders.aspx

Ventura, F., & Ploeg, J. D. (2010). Rural development: Some tentative conclusions. *Rivista di Economia Agraria*, *65*(2), 319–335.

Visentini, P. F. (2012). A dimensão político-estratégica dos BRICS: Entre a panaceia e o ceticismo [The political-strategic dimensions of the BRICS: Between panacea and skepticism]. In F. Alexandre Gusmão (Ed.), *O Brasil, os BRICS e a agenda internacional* (pp. 187–204). Brasília: Fundação Alexandre de Gusmão.

Wang, R. Y., Si, Z., Ng, C. N., & Scott, S. (2015). The transformation of trust in China's alternative food networks: Disruption, reconstruction, and development. *Ecology and Society*, Ontario, *20*(2), 1–12.

Weis, T. (2013). The meat of the global food crisis. *The Journal of Peasant Studies*, *40*(1), 65–85.

Wesz, V. J. (2014). *O mercado da soja e as relações de troca entre produtores rurais e empresas no Sudeste do Mato Grosso (Brasil)* [The soybean market and the relations of exchange between rural producers and companies in the Southeast of Mato Grosso (Brazil)] (Unpublished doctoral thesis). Rio de Janeiro: Sociedade e Agricultura (CPDA), Universidade Federal Rural do Rio de Janeiro.

Wesz, V. J. (2016). Strategies and hybrid dynamics of soy transnational companies in the Southern Cone. *Journal of Peasant Studies, 43*(2), 286–312.

Wilkinson, J. (2009). The globalization of agribusiness and developing world food systems. *Monthly Review, 61* (4). Retrieved from https://monthlyreview.org/2009/09/01/globalization-of-agribusiness-and-developing-world-food-systems/

Wilkinson, J., & Wesz, V. J. (2013). Underlying issues in the emergence of China and Brazil as major global players in the new South–South trade and investment axis. *International Journal of Technology Management and Sustainable Development, 12*(3), 245–260.

Wilkinson, J., Reydon, B., & Di Sabbato, A. (2013). Concentration and foreign ownership of land in Brazil in the context of global land grabbing. *Canadian Journal of Development Studies, 33*(4), 417–438.

Wolf, E. (1984). *Guerras camponesas no século XX* [Peasant wars of the XXI century]. São Paulo: Global Editora.

World Bank. (2015). *Indicators.* Washington, DC. Retrieved October 7, 2015, from http://data.worldbank.org/indicator

Yan, H., Chen, Y., & Ku, H. B. (2016). China's soybean crisis: The logic of modernization and its discontents. *Journal of Peasant Studies, 43*(2), 373–395.

Ye, J., Rao, J., & Wu, H. (2010). 'Crossing the river by feeling the stones': Rural development in China. *Rivista di Economia Agraria, 65*(2), 261–294.

Yeh, E. T., O'Brien, K. J., & Ye, J. (2013). Rural politics in contemporary China. *Journal of Peasant Studies, 40* (6), 915–928.

Zhao, S. (2010). The China model: Can it replace the Western model of modernization? *Journal of Contemporary China, 19*(65), 419–436.

Chinese land grabs in Brazil? Sinophobia and foreign investments in Brazilian soybean agribusiness

Gustavo de L. T. Oliveira

ABSTRACT

Chinese companies were singled out among major investors seeking farmland in Brazil, but my own and other emerging research reveals that China still lags far behind investors from the Global North, and there is evidence that the differences between them are far less significant than was presumed. Why then have Chinese agribusinesses been singled out, even as the size and amount of their investments in Brazil – particularly when compared with those form the US, EU, Argentina, and Japan – are in fact relatively small? Who are the actors in Brazil that have contributed to this apparent sinophobia, and who has challenged it? Who benefits? And how have Chinese investments themselves been affected by this disproportionate negative attention? I argue that challenges to national and food sovereignty arise, ultimately, from the transnational soybean production system regardless of the national character of any particular companies or their cross-border relations.

Introduction

Brazil–China commercial relations increased from an almost insignificant amount before the 1990s to one of the largest bilateral trade flows in the world today: Brazil is China's leading commercial partner in Latin America, and since 2009 China surpassed the US to become Brazil's number one trade partner worldwide. This trade is highly uneven, with a broad array of Chinese manufactured exports balanced almost entirely by petroleum, iron ore, and a few agricultural commodities exported from Brazil, particularly soybeans (Acioly, Pinto, & Cintra, 2010). Currently China absorbs slightly over one-third of total international soybean trade, drawing half of this amount from Brazil, where it also represents about half of total production (Oliveira & Schneider, 2016). Consequently, Chinese agribusinesses began seeking investments in Brazil to wrest greater control over the flows and profits of the international soybean trade from North Atlantic-based transnational companies that still dominate this market. While some promote this as positive 'South–South cooperation' (H. Oliveira, 2010; Zou, Long, & Hu, 2010), many others condemn it as neocolonial 'land grabs' that displace peasants, cause environmental degradation, and deindustrialize the Brazilian economy (Grain, 2008; Jenkins & Barbosa, 2012). Although the recent global land rush is more polycentric than traditional North–South dynamics (Margulis, McKeon, & Borras, 2013), it is particularly striking how the Chinese were singled out as major investors that 'bought up Africa and are now trying to buy Brazil' (Delfim Neto, quoted in O Estado de São Paulo, 2010), while other foreign investors – particularly from the US, EU, Japan, and Argentina – have been almost ignored in mainstream discourse and media reports about land grabs/foreignization of land in Brazil, especially in soybean production chains.

There is very limited fieldwork-based research on Chinese investments abroad (Armony & Strauss, 2012; Smaller, Wei, & Liu, 2012). But public sources reveal that Chinese companies still lag far behind investors from the Global North, and there is evidence that the differences between Chinese and other foreign investors are far less significant than has been presumed (Goetz, 2015; Hofman & Ho, 2012). So we must ask, why have Chinese agribusinesses been singled out for concern over land grabbing in Brazil, since the size and amount of their investments – particularly when compared with US, EU, Argentinian, and Japanese investments – are in fact relatively small? Who are the actors in Brazil who have contributed to this apparent sinophobia, and who are the actors who have challenged it? Who benefits? And how have Chinese investments themselves been affected by the legal restrictions on foreign investments in farmland, and the disproportionate concern over their entrance into the soybean complex in Brazil?

Drawing on agrarian political economy as my theoretical framework (Bernstein, 2010), rather than 'discourse analysis' as it is often understood, my analysis is assembled by the triangulation of (1) dozens of extended interviews and informal conversations with key corporate and government officials in both China and Brazil between 2011 and 2015, (2) field site inspections over a nonconsecutive period of 27 months, (3) government, company, and think-tank reports and documents related to the projects discussed, (4) extensive research of media sources and local archives, and (5) the relevant secondary literature.

After a brief review of public data and relevant literature, I demonstrate that Chinese investments in Brazilian soybean agribusiness are smaller than those from the Global North, especially in farmland, and therefore the (generally negative) attention they received in Brazilian and international media is disproportionate. This was first triggered by social movement resistance to foreign land grabs, yet their anti-agribusiness interests became subordinated in Brazil to a political compromise between the leading Workers' Party and the landed and agribusiness elite whereby government officials make it seem that they are undertaking measures to protect unspecified domestic interests from foreign land grabs, while domestic and more established agribusinesses from the Global North continue to expand control over farmland and agroindustry. This dissimulation was accomplished through the strategic use of biased media reports by an alliance of Brazilian large-scale landowners, industrialists, free-market economists and lawyers to add pressure against Workers' Party governments, generate fear and restrictions that would disproportionately affect Chinese investors in Brazil, and position themselves as necessary partners for Chinese and other foreign investors seeking Brazilian farmland. Ultimately, I argue this has been part of a broader effort by US- and European-based agribusinesses, financiers, and government-linked intellectuals to simultaneously shift attention away from the role of the Global North in transnational land grabs, and suppress rising Chinese competition against them in international agroindustrial markets. Nevertheless, I do not condone (the relatively few and smaller) Chinese land acquisitions in Brazil, as land grabs deserve critical assessment regardless of their origin. My argument demonstrates instead that challenges to national and food sovereignty arise, ultimately, from the transnational soybean production system regardless of the national character of any particular companies or their cross-border relations.

Public data and literature review

The Central Bank of Brazil (BC) is the only institution that holds definite data on foreign investments in Brazil, but legal requirements of 'financial secrecy' prohibit it from making specific company data public. Therefore, its reports only provide an overview of capital influx by national origin and broad-sector of destination, demonstrating an increasing flux of Chinese direct investments in Brazil (from 45th place in 2000 to 15th place in 2010), but a very small participation in agribusiness. In fact, between 2010 and 2013 when Chinese agribusiness investors were thought to be 'invading' Brazil (see discussion below), it is very evident that investments from the Global North dwarfed Chinese investments (Table 1).

Since BC reports do not account of the ultimate source of large flows through tax havens and combine investments in the production of all agricultural commodities, but on the other hand exclude investments in agroindustrial processing and logistics infrastructure, I analysed more fine-grained data specifically in the soybean sector and across its entire production chain drawing on the National Network of Information about Investments (RENAI) of the Brazilian Ministry of Development, Industry, and Foreign Trade. This databank only captures publicly *announced* investments, so it serves only as starting point for further multi-method research that identifies other deals and confirms announcements through fieldwork. Still, RENAI data also suggest that Chinese companies lag far behind investments from the US, EU, and Japan in the Brazilian soybean complex (Table 2; RENAI, 2012). Data from the originating side of these investments reveal a similar situation, and a more detailed discussion of reports from the Chinese Government, as well as an analysis of the methodological challenges involved with all this data is provided elsewhere (Oliveira, 2015).

Government data from both countries show no evidence of large-scale Chinese agribusiness investments in Brazil, but given the widespread sense that such databanks are not reliable and that negotiations and investments are undertaken secretly, academic literature includes widely

Table 1. Top 29 foreign investors in Brazilian agriculture: 2010–2013 (in million USD).

Origin	2010	2011	2012	2013	Total
US	1056	1058	2183	3275	7572
UK	154	1091	1510	207	2962
Luxemburg	790	1044	389	511	2734
Switzerland	373	358	586	1377	2694
Chile	537	709	633	561	2440
France	472	553	664	664	2353
Netherlands	319	280	304	576	1479
Br. Virgin Islands	277	224	201	193	895
Panama	187	160	168	164	679
Portugal	156	133	134	130	553
Japan	70	189	157	136	552
Italy	145	125	126	123	519
Denmark	139	119	120	117	495
Argentina	137	109	121	131	498
Canada	176	119	43	74	412
Germany	93	92	92	90	367
Jersey	0	0	0	171	171
Uruguay	45	38	39	38	160
Australia	21	19	38	40	118
Bermuda	8	7	7	49	71
India	0	19	19	18	56
South Korea	7	6	6	6	25
China	6	5	5	5	21
Cayman Islands	6	5	5	5	21
Spain	4	3	3	3	13
Colombia	3	3	0	0	6
Isle of Man	0	2	2	0	4
Mexico	1	1	1	1	4
Norway	0	1	0	0	1

Source: Elaborated by the author from Central Bank of Brazil, Foreign Capital Census.

Table 2. RENAI data on foreign investment announcements in the Brazilian soybean complex: 2009 to mid-2015 (millions of USD, rounded to the nearest million).

Origin	Seed and soy production[a]	Soybean processing[b]	Silos and warehouses[c]	Port and railroad[c]	Total
Japan	7	–	3.932		3.939
France	30	9	667	2.000	2.706
Netherlands	–	36	–	2.463	2.499
Algeria	–		2.150		2.150
USA	959	389	30	96	1.474
China		300	27 (2.100 cited in 2011, retracted in 2012)		327
Hong Kong	–	200	–	–	200
UAE	–	–	–	171	171
Russia	–	–	117	–	117
Portugal	–	58	–	–	58
Argentina	26	–	–	–	26

[a]Acquisition of farmland, equipment, and other production costs cited by companies that have soybeans as one of their main products.
[b]Soybean crushing and biodiesel production facilities when soybean is cited as the main feedstock, and their directly associated silos and warehouses.
[c]Total investment when soybeans are among the main commodities stored/traded through this infrastructure.
Source: Elaborated by the author from RENAI 2009 to 2015.

divergent estimates based upon media reports. Some of the more alarmist accounts estimate an exorbitant seven million hectares were grabbed in the immediate aftermath of the global financial crisis, an area the size of Ireland (Acioly et al., 2010; Faleiros et al., 2014). Yet more sober assessments from both Chinese and Brazilian scholars who attempt to investigate specific deals only identify about a handful of companies and projects that fall far short of the extensive land grabs feared (Guo, 2013; Ma & Tian, 2015; Moreira, Bonolo, & Targino, 2013; Wilkinson & Wesz, 2013; Zhou, Liu, & Guo, 2011), and those that did report multi-billion dollar investments in 2010 (e.g. CEBC, 2011) revised down their estimates significantly in the following years (CEBC, 2013). In fact, one of the few qualitative case studies on the topic illustrated how both multi-billion dollar negotiations dominating headlines in 2010 were halted altogether two years later (Lucena & Bennett, 2013).

Chinese and other foreign investments in Brazilian soybean agribusiness

So what Chinese investments are actually taking place in Brazilian soybean agribusiness, and how do they compare with other foreign investments?[1] In seeds and agrochemicals, the predominance of companies from the Global North was virtually undisputed until February 2016. Monsanto, Syngenta, and DuPont/Pioneer control over 55% of global soybean seed markets, and this concentration is even greater in South America where GM varieties predominate. Syngenta, Bayer, Basf, Dow, and Monsanto control 69.5% of global agrochemical markets, and the first three alone control over 49.1% of the USD $11.5 billion agrochemical market in Brazil (EcoNexus, 2013; Silva & Costa, 2012). Starting in the late 2000s, a few Chinese companies began registering themselves in Brazil to obtain permits for commercializing their agrochemicals directly, such as Shandong Rainbow Chemical and the Ningbo-based Tide Group. But this process takes several years and provides no guarantee of market share in Brazil, so some companies pursue more aggressive strategies through mergers and acquisitions (M&As). Nutrichem/Chongqing Huapont Pharm purchased a 7.5% stake in in CCAB Agro in 2012, a consortium of large-scale soy farmer cooperatives structured to acquire large volumes of fertilizer and agrochemicals at lower prices due to economies of scale. Yet the partnership collapsed in 2015 and Nutrichem is currently establishing its own independent operations in Brazil.[2] Notably, all these investments remain significantly smaller than the new projects announced by the major North Atlantic-based seed and agrochemical companies in Brazil (RENAI, 2009–2015).

ChemChina became a major exception in February, 2016, when it announced the purchase of Syngenta for $43 billion USD.

ChemChina had already acquired the Israeli chemical company Makhteshim Agan Industries (now rebranded as Adama) in 2011, thereby gaining control over two agrochemical factories in Brazil and their commercial licences and marketing channels. Its proposed acquisition of Syngenta is still under regulatory examination, and if successful, it will radically restructure the role of Chinese agribusiness in the seed and agrochemical sectors around the world. Although led by ChemChina, there is evidence that several other Chinese seed and agrochemical companies may expand further in Brazil in the near future, particularly as they already control supply of precursor chemicals to companies from the Global North (Oliveira & Schneider, 2016). In neighbouring Bolivia, for example, Chinese investors were already coming neck-to-neck with the leading agrochemical companies from the Global North even before ChemChina's acquisition of Syngenta (McKay, 2015). Yet it is important to note this major acquisition has taken place well *after* the imagined 'invasion' of Brazil by Chinese investors (2009–2013), and through a global-level negotiation that largely sidesteps its perception as a Chinese investment, albeit indirectly, in Brazilian agribusiness.

In agricultural production itself, foreign investments are very strongly associated with the sugarcane and eucalyptus sectors in Brazil, where there is greater incentive to guarantee supply and set local prices (surrounding sugar mills) and provide greater juridical security for long-term (5–7 year) investments in tree plantations (Wilkinson, Reydon, & Di Sabbato, 2012). The soybean and grain production sectors, on the other hand, provide less incentive for farmland ownership, which may in fact be outweighed by the political and environmental challenges for foreign acquisition of extensive tracts of farmland. Thus, the strongest presence of foreign capital in soybean production within Brazil comes through specialized farm management companies (*pools de siembra*). It may certainly be the case that some Chinese capital has been channelled through such means (cf. Faleiros et al., 2014), but the only case I have been able to confirm is the Hong Kong-based Pacific Century Group's minority-stake participation in CalyxAgro's 27,397 ha. In contrast, investors from the Global North and new entrants from Argentina and Japan have increased their participation at a far larger scale. Since 2008, leading farm management companies such as Cresud/Brasilagro, Adecoagro, SLC Agrícola, El Tejar, TIAA-CREF, Multigrain/Xingu Agro, and V-Agro acquired collectively over 750,000 in Brazil, drawing venture capital from multiple financial partners in Brazil, EU, and especially US and Japan (Oliveira, 2015; Oliveira & Hecht, 2016).

The only direct investments by Chinese companies in soy production confirmed through fieldwork are relatively small and very troubled. The first was a partnership by the private Zhejiang Fudi Agriculture Company with the Beidahuang/Heilongjiang State Farm Company. They acquired a relatively small farm in Rio Grande do Sul (around 600 ha) and then a larger one (around 16,000 ha) in Tocantins state from late 2007 to early 2008. They encountered significantly greater challenges than expected in operating the farms, so they sold the majority stake of the Brazilian subsidiary to the Chongqing Grain Group in 2011 and shifted their focus to agricultural commodity imports and agricultural production investments within China itself.[3] The Chongqing Grain Group (hereafter CGG) has been the most prominent case discussed in journalistic and academic references, as they intended to purchase 200,000 ha to farm soy in western Bahia state. At the moment CGG was about to make its purchases, the Brazilian Government imposed greater restrictions on foreign acquisition of farmland.[4] Given the juridical uncertainties of the moment, CGG opted instead to purchase a relatively smaller but still large-scale farm (around 52,000 ha), and construct a soybean processing facility to garner Brazilian support. This agroindustrial project gathered much attention when it was announced and also when it became evident it was scrapped (Staufer, 2014), but CGG never publicly acknowledge this cancellation or its *de facto* ownership of farmland in Brazil. Legal ownership of the farm rests at the hands of Brazilian figureheads, but field site inspections confirm that all three farms are *de facto* owned and operated by CGG's Brazilian subsidiary Universo Verde Agronegócios, which has faced significant losses, an environmental lawsuit, managerial problems across its farming operations in Brazil, and an occupation of its smallest farm by the Landless Rural Workers' Movement (MST) denouncing its 'foreignization' and abandonment by the Chinese owners, who left the land fallow since 2013.[5]

Another high-profile set of negotiations that were characterized as Chinese land grabbing involved the private Sanhe Hopeful Grain and Oil Co., and the central government-owned China National Agriculture Development Group (CNADG), which exchanged visits with Goiás state government and agribusiness officials during 2009 and 2010. CNADG quickly abandoned negotiations after the legal restrictions on acquisition of farmland by foreigners in 2010, and the *Folha de S. Paulo* newspaper reported that Sanhe Hopeful was planning a $7.5 billion USD investment in warehouses and farmer finance in Goiás over the next 10 years, expecting in exchange direct purchases of six million tons of soybeans (Maisonnave & Carazzai, 2011). This announcement was then echoed widely in the Brazilian press (Carfantan, 2012; O Repórter, 2011; Vital, 2010), and to an even greater extent in major Anglophone media outlets like the *New York Times* and the *Financial Times* (Barrionuevo, 2011; Hearn, 2012; Merco Press, 2011; Powell, 2011). Nonetheless, Sanhe Hopeful officials later explained to me they were frustrated and surprised with these news articles as they greatly exaggerated their intentions. Instead of investing billions in Goiás, in late 2010 Sanhe Hopeful spent a few million to acquire a 20% stake in the construction of a new grain terminal in Santa Catarina state (which obtained its environmental licence for construction only in 2016), and they declare that any further investments will only be considered when and if this port investment can actually materialize.[6]

No other farmland and greenfield (i.e. new) agroindustrial projects by Chinese companies have been confirmed through access to government documents, interviews, and other fieldwork methods, and in fact there is some evidence that at least some of the companies frequently cited as pursuing farmland investments were mere speculative adventures and overblown statements based upon initial contacts. The group Pallas International, for example, supposedly signed a memorandum of understanding (MoU) with the Bahia state government for investing in 250,000 ha for agricultural production, but government officials denied having any knowledge of such negotiations and there are no records of this MoU in government archives.[7]

Moving from soybean production to processing and trade, the sector remains marked by the continued oligopoly of the trading companies from the US and EU, particularly ADM, Bunge, Cargill, and Louis Dreyfus, collectively known as ABCD (Wesz, 2016). Since 2008, these have launched major investments to control new routes for soybean commercialization through the Amazon basin, and strengthened as well their purchasing operations throughout Brazil's soy growing areas with new and expanded warehouses. New players among commodity traders also moved into the sector, such as the European mineral commodity giant Glencore with its acquisition a soy crushing facility in 2009 and a joint-venture with ADM for the expansion of a soybean port in 2015. Glencore then turned 40% of its agribusiness subsidiary to the Canada Pension Plan Investment Board in 2016. Most impressive of all, however, have been the M&As of Japanese *sogo shosha* (general trading and industrial conglomerates). Afraid to lose their market shares in China and the rest of East Asia, they also began acquiring supply-side soybean companies. Mitsubishi, Mitsui, Marubeni, Toyota, and Sojitz, to name only the most prominent, have gained strong footholds in Brazilian soybean production and origination through their acquisitions of the Brazilian operations of Los Grobo,

Multigrain, Gavilon, Nova Agri, and Catagalo, respectively (for more detail see Hall, 2015; Oliveira, 2015; Oliveira & Hecht, 2016).

Some Chinese companies also pursued such M&As. Since 2003, Chinatex established a grain trading office in São Paulo and close ties with soy farmer cooperatives in Rio Grande do Sul state, and sought investments throughout the soybean production chain. But complications during its first large-scale shipments of soy in 2004 halted its projects, and several attempted negotiations for M&As with medium-size domestic and transnational trading companies operating in Brazil collapsed before reaching an agreement.[8] Currently, the China National Cereals, Oils and Foodstuffs Co. (COFCO) has replaced Chinatex as the main vehicle among central-government SOEs to launch large-scale foreign agribusiness investments. In early 2014, COFCO acquired majority stakes in the Dutch-based seed and trading company Nidera, and the agribusiness arm of the Singapore-listed and Hong Kong-based Noble Group for an estimated $2.8 billion USD.[9] They were the fastest rising soy and grain trading companies in Brazil from 2009 to 2014, and both have significant assets in several other countries where agricultural commodities are produced and received as well. Although COFCO does control about 145,000 ha in Brazil associated with the four sugar mills it acquired through Noble, it is very clear they intend to focus on agricultural commodity trading instead of production.[10] These M&As represent a significant step by COFCO to transform the global agroindustrial landscape in its favour, since it is already China's largest food processor and trader, and now – together with ChemChina's acquisition of Syngenta – they account for the largest and most significant presence of Chinese agribusiness capital not only in the Brazilian soy complex, but in its agribusiness sector as a whole (Murphy, Burch, & Clapp, 2012; Oliveira & Schneider, 2016). This strategy of pursuing M&As instead of direct greenfield investments has continued to yield positive results for Chinese companies interested in the Brazilian soy complex. Most recently (April 29, 2016), the Hunan Dakang Pasture Farming Co., a subsidiary of the powerful Shanghai Pengxin Group, announced the purchase of a majority stake in Fiagril, a medium-sized Brazilian soy trader and processor. The deal gives the Chinese investors control over warehouses and a soy crushing facility, but excludes Fiagril's shipping, biofuel, and seed production business.

This review demonstrates that Chinese investments in Brazilian soybean agribusiness are not in fact focused on farmland, where traditional investors from the Global North far outpace Chinese newcomers. In order to properly grasp the entrance of Chinese investments in Brazilian agribusiness, therefore, we must broaden the focus from land to other links in agribusiness production chains, and from new projects and direct investments to M&As. Doing so reveals that the leading Chinese soybean agribusinesses display goals and *modus operandi* remarkably similar to the leading companies from the Global North. So rather than taking the discourse of China as a 'land grabber' for granted, we must investigate how it has been produced internationally and in Brazil.

China in the 'global land grab' discourse

It is clear that Chinese agribusinesses are new entrants facing great challenges with greenfield investments in the Brazilian soybean complex. They have only established a strong presence in seed/agrochemicals and processing/trading through M&As led by ChemChina and COFCO, but beyond that Chinese investments are dwarfed by those from the US, Europe, Argentina, and Japan, particularly regarding farmland itself. Why, therefore, have Chinese agribusinesses been singled out for concern over land grabbing in Brazil?

This is partially because China is a *new* large-scale investor competing against established agribusinesses and seeking to cut-off middlemen from the Global North. However, this alone cannot explain the sinophobia witnessed, since similar reaction should have been expected towards other new entrants, Argentinian and Japanese companies being the most prominent among them. Argentina may have escaped alarm since it is a familiar neighbour, and while Japan did cause similar concern when it began to finance the expansion of soy production into the Brazilian Cerrado region during the 1970s and 1980s (Martins & Pelegrini, 1984), the close ties it has since established with Brazilian agribusinesses may have allowed its large-scale investments in soybean production and export infrastructure in Brazil to fly under the radar. There must be something specific to Chinese investments, therefore, that explains why and how China was singled out for concern.

The fact their investments in 2010 went primarily to petroleum production in Brazil (CEBC, 2011) made it easy to characterize China as a 'resource seeking' new player that operates differently than the 'commercially interested' investors from the Global North (O Estado de São Paulo, 2010). Moreover, since investors from the Global North have been in Brazil for a longer time and are defending their hegemonic position in the market, they can control new investments and M&As more easily than new entrants. And surely, the Chinese investments that went through most smoothly are precisely the ones structured as partnerships with Brazilian and other agribusiness companies already operating there. If some Chinese companies – especially state-owned enterprises – became target of so much negative attention, it is in part because their lack of experience, lack of connections, and political *modus operandi* made them rely on Brazilian Government officials who gave a high-profile to their negotiations, instead of striking low-profile deals directly with private agribusiness companies.

This also appears to have been the case with the failed Iranian attempt to purchase or lease large-scale farms in Brazil during 2009 (Reuters, 2009). However, if this was the main factor causing the great furore witnessed about 'Chinese land grabs' in Brazil, we would expect similar reactions to other new entrants in similar conditions. Yet this did not take place with the Iranian case just mentioned, an Indian purchase of over 106,000ha during 2010 (Shree Renuka, n.d.), Saudi and Qatari negotiations that same year (Qatar News Agency, 2010; The Peninsula, 2010), or the massive Algerian investments announced in processing, warehouses, and ports for exporting soy (RENAI, 2015). In order to properly understand why Chinese investors have been singled out for such concern in Brazil, it is necessary to investigate who has contributed to this apparent *sinophobia*, who has challenged it, and who ultimately benefits.

Although the Chinese Government's policy of 'going out' was launched in 2001 and agricultural investments were an integral part of this government strategy from the get-go, the 'China as land grabber' discourse only gained prominence with the signing of a MoU in January 2007 for leasing at least 1.4 million acres in the Philippines right when a food-price crisis began to be felt sharply in that country and elsewhere. Filipino and Anglophone media reports painted a very negative image of the deal, suggesting that it was intended to export rice to China even while the poor in the Philippines went hungry (e.g. IBON Media, 2008; The Economist, 2009), and this argument was reproduced in spades by the Texas-based corporate intelligence agency Stratfor, describing how 'Beijing' was pushing its agribusinesses to 'outsource farming projects overseas' to guarantee the country's food security, and situating this as part of a broader threat of 'neo-imperialism' through resource-seeking investments (Stratfor, 2008).

The 'China as land grabber' discourse was clearly coming from the Global North, and this continued especially through the work of UK- and US-Government-affiliated researchers concerned about the impact that China's foreign investments might have for Anglo-American interests around the world (Ellis, 2012; Salidjanova, 2011; UKTI, 2014), as well as US agribusiness think tanks (Soy and Corn Advisor, 2010) and agribusiness-financing institutions from the Global North such as the

Rabobank (Rasmussen, Chow, Nelson, Hendricks, & Sarvanti, 2011). However, the 'China as land grabber' discourse also resonated strongly with peasant and working class resistance to agribusiness domination and spiking food prices in the Philippines: peasants led waves of massive protests and the Chinese deal was ultimately not consummated, but the stigma of 'land grabber' stuck to China as more and more media and think-tank reports sought to implicate it in what was beginning to be discussed as a 'global land rush' (Asian Peasant Coalition, 2009; Marks, 2008; von Braun & Meinzen-Dick, 2009).

This 'global land grab' discourse was most prominently articulated by the GRAIN report *Seized! The 2008 land grab for food and financial security* (GRAIN, 2008), which identified a fast and massive flow of sovereign wealth, financial, and agroindustrial capital into farmland and agricultural commodities due to the conjuncture of a decade-long commodity boom culminating with a sharp spike in food prices in 2007, followed by the pronounced financial crisis that spread from Wall Street around the world in 2008. Although this report correctly cited financiers from the Global North shifting from the uncertainty of financial markets into farmland as a key aspect of the ongoing global land grab, it focused much more on high-profile negotiations and investment announcements from China and the Persian Gulf states (and to a lesser extent South Korea and Japan), characterizing this process as being led primarily by land-poor, capital-rich, net food-importing countries launching investments in Sub-Saharan Africa.

This presumed leading role of China and other 'land-poor/capital-rich' countries became very prominently marked by the South Korean Daewoo corporation's proposal to lease three million acres in Madagascar, which received especially bad press in Anglophone media, sparked massive protests by Malagasy peasants and urban workers, and ultimately led to the collapse of the Malagasy Government and Daewoo's proposal (Burgis & Blas, 2009). At the point that land grabbing became implicated in the downfall of a government and the world economy was falling in a downward spiral triggered by the financial crisis in the US, this discourse – and China's prominent role therein – reached the upper echelons of all governments and multilateral institutions, Brazil included.

Underlying concerns over 'foreignization of land' in Brazil

Concern over 'foreignization of land'[11] had already become integral to political debates in Brazil by 2007, when the most prominent example discussed was Stora Enso's operations in Brazil's southern border (a Swedish-Finish forestry and cellulose company), which was critiqued by the Landless Rural Workers' Movement (MST), attacked by the women of La Via Campesina in southern Brazil, and cited as a clear example of a foreign corporation taking advantage of weakened restrictions on acquisition of farmland (Estado, 2008). The president of the National Institute for Colonization and Agrarian Reform (INCRA), the government agency responsible for regulating foreign farmland investments until 1995, pressured by the media to disclose how much land was actually controlled by foreigners in Brazil, admitted the government had lost the ability to track and regulate this for over a decade, and rode the wave created by peasant movements to demand the reinstatement of monitoring and restrictions on foreign acquisitions of farmland (Roldão, 2007, 2008b). The sugar/ethanol and the forestry/cellulose industries lobbied hard against a revision of the regulations, threatening to cancel and even withdraw investments from Brazil. Given the desire by the federal government to maintain all the economic activity possible while the financial crisis was blowing up in 2008, the revision of regulations was stalled and a judicial committee was created to research and make recommendations for future decision.

Until 2010, however, Chinese investors were not yet the 'boogey men' of the land grab discourse in Brazil, which turned around concerns that foreign capital flowing into Brazilian farmland was merely speculative, associated with money laundering, a threat to national sovereignty (especially in border areas and the Amazon), and/or associated with the subordination of Brazilian natural resources to exploitation by the traditional and hegemonic agribusiness and extractive interests from the Global North. The discussion in academic research and the leading newspapers during 2008 and 2009 about the uncertain amount of farmland under foreign control (and the presumption that official INCRA data significantly underestimate the actual hectarage) made little or no mention of Chinese companies or individuals as large-scale landowners (e.g. Folha Online, 2008; A. Oliveira, 2010; Pretto, 2009; Roldão, 2007, 2008a; Valor, 2009). As late as November 2010, an alarmist head-line in one of Brazil's leading newspapers claimed 'Foreigners buy 22 soccer fields per hour' (Odilla, 2010), but made no mention of Chinese investors whatsoever.

Chinese land grabbers and scapegoats

The year of 2010 was pivotal for the manner in which China was brought into the 'foreignization/ land grab' discourse in Brazil. A highly anticipated report by the World Bank released that year implicated China as first on the list of the main players that characterize the current land rush as distinct from previous waves of large-scale agricultural investments around the world, even while it recognized the frailty of evidence and possibility of bias from media reports, and the undeniable role of Anglo-American capital:

> Press reports allow identification of source countries without complicated searches in the company reg-istry. Although part of this may reflect *reporting bias* or *strategic use of press reports by some types of investors*, most of the projects in the [GRAIN] database originate from a few countries. These include *China*, the Gulf States ... North Africa ... Russia, and such developed economies as the United Kingdom and the United States. (World Bank, 2010: p. 53, emphases added)

The World Bank's report – and its identification of China's leading role in the global land grab – resonated with leftist intellectuals in Brazil (Fernandes, 2011; Sauer, 2010). It was also at this moment that Chinese investments in Brazil jumpstarted in the oil and energy sectors, and Chinese delegations to Brazil seeking farmland and agroindustrial investments increased from sporadic and piecemeal efforts to a high-profile and steady flow. These delegations were particularly notable due to the involve-ment of state-owned companies like CNADG, COFCO, and CGG, government delegations from the provincial level up to the Minister of Agriculture, and mentions of additional companies the identity of which remained undisclosed (Chade, 2010; Inácio, 2010; Jornal Opção, 2010; Portal, 2010). Members of the presidential cabinet who had already requested the Attorney General to review and strengthen restrictions on acquisition of farmland by foreigners earlier in 2008 then coor-dinated with the Federal Public Ministry to give the most gravity possible to the judicial commis-sion's recommendation in favour of stricter regulations, decreed on August, 2010.[12]

At this time, however, even though leftist academics and social movements remained opposed to 'foreignization' of land, the high-level political orchestration around the new restrictions also involved clear coordination with certain sectors of Brazilian large-scale landowners and agribusiness companies. This triggered a critique of the president of INCRA and some high-level Workers' Party officials for orchestrating a 'farce' to distract leftist intellectuals and social movements from the reversal in the government's land redistribution policy in favour of land regularization in the Amazon-Cerrado transition zone, and fool them into supporting what had by then become a set of legal restrictions that would neither effectively stop foreign investments in Brazilian farmland, nor weaken the power of Brazilian agribusiness and large-scale landowners, but rather consolidate the power of the latter as necessary partners and brokers for foreign investors (A. Oliveira, 2010; Oliveira, 2013; Scoton & Trentini, 2011).

Some legal experts deeply involved with Chinese and other foreign companies seeking agroindustrial investments in Brazil even suggest that the timing selected by the Lula administration to announce the new restrictions demonstrates how it was intended to obtain the support from key sectors of the landed and agribusiness elite for the election of Dilma Rousseff, his picked successor from the Workers' Party.[13] Formal positions taken by various agribusiness associations reflect a divided class, where those from areas where middle- and large-scale landowners operating primarily with soy production and ranching supported restrictions to limit foreign competition with their own expansion and place themselves as necessary partners to foreign investors (e.g. the Agriculture and Livestock Federation of Mato Grosso – FAMATO, and the Association of Irrigation and Farmers of Bahia – AIBA), while opposition came from landowner associations from regions where the sugar/ethanol industry and the forestry/cellulose industry were expected to lead agribusiness expansion (e.g. the Agriculture and Livestock Federation of Mato Grosso do Sul – FAMASUL) and the main representatives of those specific agroindustrial sectors. In order to attend these multiple and conflicting interests in the aftermath of the restriction on foreign acquisition of farmland, the congressional proceedings on the drafting of a new law on the matter had to find a compromise. The final report of the congressman who launched the proceedings, Beto Faro of the Workers' Party, sustained the restrictions to avoiding land speculation and loss of national sovereignty. But the president of the congressional commission on agriculture at the time, Homero Pereira from the Republican Party of Mato Grosso, who was also the leader of the landowner and rancher bloc in congress, replaced the final report with an alternate by Marcos Montes, a staunchly pro-agribusiness congressman of the Democratic Party from Minas Gerais, loosening restrictions on all but sovereign wealth funds and state-owned companies – a barely veiled way to make the Chinese investors become the perfect scapegoats.

Sinophobia – Who promotes? Who challenges? Who benefits?

Taking advantage of the naïve fanfare that the Chongqing Grain Group was bringing to Brazilian, Chinese, and international media alongside the Workers' Party government of Bahia in late 2010 (e.g. Decimo, 2011), a landed elite and industrialist alliance launched a massive campaign in Brazilian media to demonize Chinese investors as state-backed, politically interested land grabbers, and consequently the primary target and justification for the tighter restrictions imposed on foreign acquisition of farmland (Carfantan, 2012; Cruz & Vaz, 2011; O Estado de São Paulo, 2010; Valor, 2010). The domestic agribusiness and landed elite were represented primarily by congressman Homero Pereira, while the industrialist bloc had Delfim Neto (the influential economist and past Minister of Finance) as their leading spokesperson. When I interviewed congressman Pereira in 2012 about the ongoing congressional proceedings, he immediately and insistently brought up the 'threat' of Chinese land grabbing, even though I had never mentioned to him that Chinese investments were also the focus of my research.[14] Delfim Neto – author of the famous expression that 'the Chinese have bought up Africa and are now trying to buy Brazil' – has longstanding ties to the São Paulo-based industrial elite that was seeking anti-dumping measures and other sorts of government supports for the manufacturing sectors of Brazil that are being outcompeted by cheaper Chinese imports. They certainly understood that stoking sinophobia in the agribusiness sector could also strengthen their case for the anti-Chinese commercial policies they desired (Cunha, 2011). And both sides were also united in their desire to weaken the Workers' Party government in Bahia, the main site of CGG investments and a previous bastion of pro-agribusiness and industrialist politicians that had just been lost to the left in 2007.

At the peak of this campaign for sinophobia, the right-wing *Época* magazine published an extensive account of Chinese investments in Brazil entitled 'The Chinese Invasion', estimating total investments between 2010 and 2012 at over $35 billion USD, which included $17.2 billion USD for petroleum and energy (Sinopec and Stategrid) and $8.2 billion USD in agribusiness, mostly composed by the hyper-inflated $7.5 billion USD estimate for Sanhe Hopeful's 'investment' in Goiás (Todeschini & Rydlewski, 2012). Chinese investors, of course, were significantly affected by this campaign, which remains a topic of much frustration for them. The executive president of Hopeful Investment and Holdings told me with frustration simmering behind is calm demeanour: 'The news people put all the companies that have potential to do business in Brazil in a chart, and make it look like a huge investment. That is kind of misleading.'[15] Indeed, and it appears that was precisely the intention.

Ultimately, Congress failed to pass any new legislation on foreign acquisition of farmland, and even though the Attorney General's 2010 interpretation could certainly be debunked as unconstitutional by a challenge in the courts (Hage, Peixoto, & Vieira Filho, 2012), foreign agribusinesses and financiers found it more convenient to work around the restrictions with Brazilian partners (Fairbairn, 2015). Therefore, opposition to these restrictions was largely limited to those particular sectors of Brazilian and transnational agribusiness most affected, such as the sugar/ethanol and forestry/cellulose industries, which were reported to freeze $3.2 billion USD in the immediate aftermath of the Attorney General's restriction, and accounted for the bulk of the $15 to $25 billion USD in 'lost investments' estimated by the Brazilian Agency for Rural and Agribusiness Marketing in 2011 (Barros & Pessoa, 2011; Borges, 2010). Opposition to the restriction – and to the 'China as land grabber' discourse in general – also came from Chinese Government officials, investors, and brokers specialized in attracting Chinese investments to Brazil (Barrionuevo, 2011; Maisonnave & Carazzai, 2011). A Brazilian partner of the CGG who is one of the *de jure* owners of the farm they operate in Bahia went so far as to accuse Delfim Netto and a group of lawyers with ties to US-based agribusinesses of drafting the restriction for the Attorney General precisely at that moment in order to undermine the specific land acquisitions that CGG had been negotiating, and argued very convincingly that the restriction served primarily to protect agribusinesses from the Global North from the incipient competition of Sino-Brazilian agribusiness partnerships.[16] Ironically, however, as a wealthy landowner in western Bahia who became partner to one of the largest Chinese agribusiness investments in Brazil, he benefitted more directly than the ABCDs from the restriction against acquisition of farmland by foreigners, at least in the short term. Lastly, a weaker and more ideological opposition comes from free-market economists and lawyers in Brazil who see the measure as yet another example of the unnecessary and irrational protectionism of the Workers' Party government that restricts foreign investments considered to be necessary for the growth of the Brazilian economy (Hage et al., 2012; Nasser, 2010; Pombo, 2012).[17]

Finally, how have Chinese investments themselves been affected by these restrictions, and the disproportionate concern over their entrance into the soybean complex in Brazil? Clearly, those companies and individuals who envisioned farmland acquisitions as their entrance into Brazilian agribusiness were redirected to other countries, or figured out that it was more viable to pursue investments in other links of agroindustrial production chains. While it would be inaccurate to describe Sanhe Hopeful's case as a 'shift' from land-based investments in Goiás to infrastructure-based investments in Santa Catarina as some media sources have done (e.g. Sant'anna, 2014), this does capture the broader trend of Chinese investments in Brazilian agribusiness. The most representative agribusiness association of Brazil (CNA), for example, partnered with its sister organizations in the industrial and transportation sectors (CNI and CNT) and the Brazil–China Business Council (CEBC) to organize a major forum during Xi Jinping's state visit to Brasília in 2014 to inform and invite Chinese investors into Brazil's logistics infrastructure, particularly railroads that can cheapen agricultural commodity exports (Marin, 2014), and they followed up with a similar event during premier Li Keqiang's visit in 2015.[18] This reveals how sinophobia has not only been instrumentalized for the benefit of certain actors as described above, but also how it curtailed a more accurate understanding of the actual dynamics of Chinese investments in the Brazilian soybean sector: despite involving state-owned companies and government-oriented negotiations, it largely reflects the business logic of transnational capital also witnessed in the operations and priorities of agribusiness investments from the Global North and its absorption by the Brazilian state and elites.

Conclusion

This argument should not be confused with condoning Chinese state and corporate efforts – whether successful or not – at establishing large-scale agribusiness operations in the Brazilian soybean complex and agribusiness sector more generally. I also do not deny that during the first 10 years of China's 'going out' policy (2001–2011), many Chinese agribusinesses went to Brazil seeking primarily land-based investments. However, the international sinophobia that characterized the global land grab discourse and Brazilian debates on 'foreignization of land' starting around 2010, and institutionalized that year in the Attorney General's establishment of greater restrictions for acquisition of farmland by foreigners, certainly limited Chinese land-based investments that were being prospected or even negotiated at the time. This also slowed the advancement of Chinese investments in other links of the soybean production chain due to greater hesitation and lack of trust between Chinese and Brazilian partners, and resulted in the strengthening of Brazilian large-scale landowners and agribusinesses from the Global North. The convergence of their interests here, however, does not necessarily imply a harmonious relationship all around.

Some Chinese companies pursued government-oriented negotiations for farmland and largely failed, while others pursued private M&As successfully in other links of the production chain – but these different strategies do not align with state-owned vs. private capital. Against the sinophobic and simplistic claim that Chinese investments in Brazil (and elsewhere) are especially worrisome because they act upon a 'state logic' unlike the commercial interests of other investors (O Estado de São Paulo, 2010), I demonstrate that the globalization of Chinese agroindustrial capital is in fact taking place through transnational M&As designed to challenge the hegemony of oligopolies from the Global North, reflecting much more the commercial logic of the latter than the presumed political logic of farmland acquisitions. Remembering that the waning power of agroindustrial corporations from the Global North has been a legacy of Euro-American colonization and integral to their broader efforts at maintaining hegemony during the so-called Cold War era (Oliveira, 2016; Patel, 2013), this article provides further evidence that the differences between Chinese and other foreign investors in international agribusiness are far less significant than has often been presumed (Goetz, 2015; Hofman & Ho, 2012). In turn, this also indicates that to properly understand Chinese investments in Brazilian agribusiness, we must broaden the focus from land to other links in agribusiness production chains, and from direct investments in new projects to M&As. Noting these similarities with agribusinesses from the Global North, the political construction of the 'China as land grabber' discourse, and the geopolitical stakes involved, we may transcend the methodologically barren and politically misleading terms in which Chinese agribusiness investments have been predominantly discussed so far. Finally, rather than condoning the relatively few and smaller Chinese land grabs that have taken place, this research also suggests that a critique of investments in soybean agribusiness should pertain to its entire transnational agroindustrial production system, regardless of the national character of any particular companies or their cross-border operations.

Notes

1. All information in this section is derived from stock exchanges, corporate websites, media, and personal interviews. Unless otherwise noted, I used only media reports that refer to projects already operational or under construction, and completed M&As, thereby avoiding unconfirmed announcements. For more detail see Oliveira (2015).
2. Interviews with Nutrichem and CCAB Agro executives, Beijing and São Paulo, 2015.
3. Interviews with Fudi and Beidahuang Group executives, Heilongjiang, 2011 and Zhejiang, 2015.
4. These restrictions were imposed by the Legal Opinion LA-01 by the Attorney General's Office (AGU) on 23rd August 2010. It determines that restrictions previously established by Law 5.709 of 1971 remain in effect despite the 6th amendment to the 1988 Constitution that removed legal distinctions between Brazilian companies and companies registered in Brazil but controlled by foreign capital. That law basically determines that foreigners and foreign-controlled companies cannot own or lease an area greater than 25% of any municipality, and no more than 10% of any municipality can be under control of foreigners of the same nationality. For more detailed discussions of this government restriction, see A. Oliveira (2010), Hage et al. (2012), and Perrone (2013).
5. The financial losses have accrued due to the preparation costs for the crushing facility that has not been built, land clearing and soil improvement costs at the farms, and poor harvests due to inexperienced management and drought. Several key documents have been obtained from municipal, state, and the federal government agencies in Brazil to triangulate these accounts, which are discussed in greater detail in my forthcoming publications. The MST occupation of the Sol Agrícola farm in Rio Grande do Sul lasted from 19th October to 14th November 2015.
6. Interviews with Sanhe Hopeful executives in Beijing and Hebei, 2013 and 2015.
7. Interviews with Bahia state government officials, 2012, 2014, and 2015.
8. Interview with Chinatex executives, São Paulo, 2015.
9. Noble has sometimes been cited as a 'Chinese company' in discussions of transnational agribusiness investments (e.g. Wilkinson & Wesz, 2013); however, this was not an adequate characterization of the company prior to the acquisition of its agribusiness arm by COFCO.
10. Interviews with Noble and Nidera executives in Brazil, São Paulo, 2014 and 2015; and with COFCO executives in Beijing, 2013 and Hong Kong, 2015.
11. 'Foreignization of land' (*estrangeirizção de terras* in Portuguese) is the expression primarily used in the Brazilian iteration of the 'global land grab' discourse, and it is sometimes combined with the more common expression *grilagem de terras* (the falsification of land titles) that is more closely associated with historical and domestic processes related to, but distinct from, the more recent 'global land grab'. For a discussion of the intimate but complex relations between these concepts and processes in Brazil, see Zoomers (2010) and Oliveira (2013).
12. See Note 4.
13. Interview with the head of the China-desk at the firm Duarte Garcia, Caselli Guimarães, e Terra, a major law firm that has among its clients several of the largest Chinese investors in Brazil, São Paulo, 2014 and 2015.
14. Interview with congressman Homero Pereira (PR-MT), Brasília, 2012.
15. Interview, Hebei, 2015.
16. Interview with agribusiness executive who requested to remain anonymous for this account, Bahia, 2014.
17. Adopting this rationale, it appears that the post-impeachment neoliberal government of Michel Temer might lift the 2010 decree to loosen restrictions on foreign investment in farmland, but no official action has yet been taken at the time of writing.

18. Participation at the CEBC/CNT/CNI/CNA events, Brasilia, 2014 and 2015; interview with the CNT and CNA representatives in China, Brasília, 2014, Beijing and Shanghai, 2015.

Acknowledgements

I thank Xu Siyuan, Ye Jingzhong, Wang Chunyu, Ben Cousins, among other colleagues from the BRICS Initiative for Critical Agrarian Studies, as well as colleagues from the Haas Junior Scholars Fellowship 2015–2016, and the anonymous reviewers for helpful comments on earlier drafts.

Disclosure statement

No potential conflict of interest was reported by the authors.

Funding

This work was partially funded by the BRICS Initiative for Critical Agrarian Studies, Inter-American Foundation [Grassroots Development Fellowship], and at the University of California at Berkeley by the Graduate Division [Eugene Cota-Robles Fellowship], the Center for Chinese Studies [Summer Research Grant], and the Institute for International Studies [Simpson Fellowship]. The views expressed here do not necessarily reflect those of any of these institutions.

References

Acioly, L., Pinto, E. C., & Cintra, M. A. (2010). China e Brasil: oportunidades e desafios [China and Brazil: Opportunities and challenges]. In R.P.F. Leão, E.C. Pinto, & L. Acioly (Eds.), *A China na nova configuração global: impactos políticos e econômicos* [China in the new global configuration: political and economic impacts] (pp. 308–350). Brasília: IPEA.

Armony, A., & Strauss, J. (2012). From going out (zou chuqu) to arriving in (desembarco): Constructing a new field of inquiry in China–Latin America interactions. *The China Quarterly, 209*(2), 1–17.

Asian Peasant Coalition. (2009, July 25). Peasant groups launch Asia-wide actions against global land grabbing. APC Secretary-General, press release. Retrieved from http://farmlandgrab.org/post/view/6501-peasant-groups-launch-asia-wide-actions-against-global-land-grabbing

Barrionuevo, A. (2011, May 26). China's interest in farmland makes Brazil uneasy. *New York Times*.

Barros, A. M., & Pessoa, A. (2011). *Impactos Econômicos do Parecer da AGU (Advocacia Geral da União), que impõe restrições à aquisição e arrendamento de terras agrícolas por empresas brasileiras com controle do capital detido por estrangeiros* [Economic impacts of the legal opinion by the Attorney General's Office that imposes restrictions to the acquisition and lease of farmland by Brazilian companies which capital controlled by foreigners]. São Paulo: Agroconsult e MB Agro.

Bernstein, H. (2010). *Class dynamics of agrarian change*. Halifax: Fernwood.

Borges, R. (2010). A terra é de todos? [Is the land everyone's?] *Dinheiro Rural* 73, November. Retrieved from http://revistadinheirorural.terra.com.br/secao/agroeconomia/a-terra-e-de-todos

Burgis, T., & Blas, J. (2009, March 18). Madagascar scraps Daewoo farm deal. *Financial Times*. Retrieved from http://www.ft.com/intl/cms/s/0/7e133310-13ba-11de-9e32-0000779fd2ac.html#axzz3WtSLgrLJ

Carfantan, J.-Y. (2012, July 17). O jogo de Go e o agro. [The game of go and the agricultural sector]. *Agrolink*. Retrieved from http://www.agrolink.com.br/colunistas/ColunaDetalhe.aspx?CodColuna=4344

CEBC (Brazil–China Business Council). (2011). *Chinese investments in Brazil: A new phase in the China–Brazil relationship*. Rio de Janeiro: CEBC.

CEBC (Brazil–China Business Council). (2013). *Chinese investments in Brazil from 2007–2012: A review of recent trends*. Rio de Janeiro: IDB/CEBC.

Chade, J. (2010, April 27). China negocia terras para soja e milho no Brasil [China negotiates land for soy and corn in Brazil]. *O Estado de São Paulo*. Retrieved from http://economia.estadao.com.br/noticias/geral,china-negocia-terras-para-soja-e-milho-no-brasil,543390

Cruz, V., & Vaz, L. (2011, November 19). Terra para estrangeiro terá mais restrição [Land for foreigners will have even more restriction]. *Folha de S. Paulo*. Retrieved from http://www1.folha.uol.com.br/fsp/mercado/9794-terra-para-estrangeiro-tera-mais-restricao.shtml

Cunha, J. E. (2011). *O impacto da importação de produtos chineses: uma análise das estratégias empresariais brasileiras e estadunidenses (1990–2011)* [The impact of Chinese imports: An analysis of Brazilian and US business strategies (1990–2011)] (*Unpublished masters' thesis*). Catholic University of Brasília, Brasília.

Decimo, T. (2011, March 18). Grupo chinês vai investir R$ 4 bi em processamento de soja na Bahia [Chinese group will invest $4 billion reais in soybean processing in Bahia]. *O Estado de São Paulo*. Retrieved from http://economia.estadao.com.br/noticias/geral,grupo-chines-vai-investir-r-4-bi-em-processamento-de-soja-na-bahia,59210e

EcoNexus. (2013). *Agropoly: A handful of corporations control world food production*. Retrieved from http://econexus.info/publication/agropoly-handful-corporations-control-world-food-production

Ellis, E. (2012). *The expanding Chinese footprint in Latin America: New challenges for China, and dilemmas for the US*. *Asie.Visions* 49. Paris: Institut Français de Relations Internationales.

Estado, A. (2008, March 4). Mulheres da Via Campesina invadem fazenda no RS [Women of La Via Campesina occupy farm in Rio Grande do Sul state]. Retrieved from http://g1.globo.com/Noticias/Brasil/0,,MUL336355-5598,00-MULHERES+DA+VIA+CAMPESINA+INVADEM+FAZENDA+NO+RS.html

Fairbairn, M. (2015). Foreignization, financialization and land grab regulation. *Journal of Agrarian Change*, 15 (4), 581–591.

Faleiros, R., Nakatani, P., Vargas, N., Nabuco P., Gomes, H., & Trindade, R. (2014). A expansão internacional da China através da compra de terras no Brasil e no mundo [China's international expansion through land acquisitions in Brazil and around the world]. *Textos & Contextos*, 13(1), 58–73.

Fernandes, B. M. (2011). Estrangeirização de terras na nova conjuntura da questão agraria [Foreignization of land in the new conjuncture of the agrarian question]. In *Conflitos no Campo Brasil* – 2010 [Conflicts in the countryside: Brazil 2010]. Goiânia: CPT.

Folha Online. (2008, June 5). Incra investiga venda de terras na Amazonia para milionário sueco [INCRA investigates land sales in the Amazon to Swedish millionaire]. Retrieved from http://www1.folha.uol.com.br/poder/2008/06/409262-incra-investiga-venda-de-terras-na-amazonia-para-milionario-sueco.shtml

Goetz, A. (2015). How different are the UK and China? Investor countries in comparative perspective. *Canadian Journal of Development Studies/Revue Canadienne D'études du Développement*, 36(2), 179–195.

GRAIN. (2008). *Seized! The 2008 land grab for food and financial security*. Briefing. Barcelona: Grain.

Guo, J. (2013, 20 November). *China's food policy: Myths and truths*. Paper presented at the China and Latin America Agriculture Workshop, Inter-American Dialogue, Washington, DC.

Hage, F. A., Peixoto, M., & Vieira Filho, J. E. (2012). Aquisição de terras por estrangeiros no Brasil: uma avaliação jurídica e econômica [Land acquisition by foreigners in Brazil: a legal and economic evaluation]. *Textos para Discussão* 114. Brasília: Center for Studies and Research of the Senate.

Hall, D. (2015). *The role of Japan's general trading companies (Sogo shosha) in the global land grab* (Conference Paper n. 3). Chiang Mai: BRICS Initiative for Critical Agrarian Studies.

Hearn, K. (2012, February 1). China plants bitter seeds in South American farmland. *Washington Times*. Retrieved from http://www.washingtontimes.com/news/2012/feb/1/china-plants-bitter-seeds-in-south-american-farmla/?page = all#pagebreak

Hofman, I., & Ho, P. (2012). China's 'developmental outsourcing': A critical examination of Chinese global 'land grabs' discourse. *Journal of Peasant Studies, 39*(1), 1–48.

IBON Media. (2008, September 22). RP-China farm deals and local agriculture: Feast or famine? *IBON Features.* Retrieved from http://farmlandgrab.org/post/view/2517-rp-china-farm-deals-and-local-agriculture-feast-or-famine

Inácio, A. (2010, May 27). Planos da China de investir no Cerrado nordestino começam a virar realidade [China's plans to invest in the northeast's Cerrado region begin to turn a reality]. *Valor Econômico.* Retrieved from http://www.seagri.ba.gov.br/noticias/2010/05/27/planos-da-china-de-investir-no-cerrado-nordestino-come%C3%A7am-virar-realidade

Jenkins, R., & Barbosa, A. (2012). Fear for manufacturing? China and the future of industry in Brazil and Latin America. *The China Quarterly, 209*(2), 59–81.

Jornal Opção. (2010, November 8). Goiás e China firmam acordo para cooperação para agricultura [Goiás state and China sign agreement for cooperation in agriculture]. Retrieved from http://www.jornalopcao.com.br/posts/ultimas-noticias/goias-e-china-firmam-acordo-de-cooperacao-para-agricultura

Lucena, A. F., & Bennett, I. G. (2013). China in Brazil: The quest for economic power meets Brazilian strategizing. *Carta Internacional, 8*(2), 38–57.

Ma X. & Tian Z.H. (2015). 中国与巴西农业贸易和投资现状及启示 [Current situation and implications of agricultural trade and investment between China and Brazil]. *Yatai Jingji [Asia-Pacific Economic Review], 1*, 60–64.

Maisonnave, F., & Carazzai, E. H. (2011, April 3). Chineses investem na soja brasileira [Chinese invest in Brazilian soy]. *Folha de S. Paulo.* Retrieved from http://www1.folha.uol.com.br/fsp/mercado/me0304201102.htm

Margulis, M. E, McKeon, N., & Borras Jr. S. M. (2013). Land grabbing and global governance: Critical perspectives. *Globalizations, 10*(1), 1–23.

Marin, D. C. (2014, July 5). China vai investir em logística no Brasil [China will invest in logistics in Brazil]. *Exame.* Retrieved from http://exame.abril.com.br/noticia/china-vai-investir-em-logistica-no-brasil/imprimir

Marks, S. (2008). China and the great global land grab. *Emerging Powers in Africa Watch*, Issue 412. Retrieved from http://pambazuka.org/en/category/africa_china/52635

Martins, P., & Pelegrini, B. (1984). *Cerrados: uma ocupação Japonesa no campo* [Cerrados: A Japanese occupation in the countryside]. Rio de Janeiro: Editora Codecri.

McKay, B. (2015). *BRICS and MICS in Boliva's 'value'-chain agriculture* (Working Paper n. 6). Cape Town: BRICS Initiative for Critical Agrarian Studies.

Merco Press. (2011, April 11). *China plans to invest 10 billion USD in soy production and processing in Brazil.* Retrieved from http://en.mercopress.com/2011/04/11/china-plans-to-invest-10-billion-usd-in-soy-production-and-processing-in-brazil

Moreira, E. R., Bonolo, F., & Targino, I. (2013). Estrangeirização das terras: Algumas notas sobre o caso do Brasil e da Paraíba [Foreignization of land: Some notes about the case of Brazil and Paraíba state]. *Boletim Dataluta, 69*, 2–11.

Murphy, S., Burch, D., & Clapp, J. (2012). *Cereal secrets: the world's largest grain traders and global agriculture* (Oxfam Research Report). Oxford: Oxfam.

Nasser, A. M. (2010, October 20). Editorial: Terras agrícolas na alça de mira [Editorial: Farmland in the crosshairs]. *O Estado de São Paulo.* Retrieved from http://opiniao.estadao.com.br/noticias/geral,terras-agricolas-na-alca-de-mira-imp-,627212

O Estado de São Paulo. (2010, August 3). Editorial: China compra terras no Brasil [Editorial: China buys land in Brazil]. *O Estado de São Paulo.* Retrieved from http://opiniao.estadao.com.br/noticias/geral,china-compra-terras-no-brasil-imp-,589697

O Repórter. (2011, May 5). Acordo entre Goiás e China prevê investimentos de US$ 7 bilhões e incremento na produção de soja [Agreement between Goiás and China predicts 7 billion USD in investments for soybean production.] Retrieved from http://www.jornaloreporter.com.br/post/676/empresas-negocios/acordo-entre-goias-e-china-preve-investimentos-de-us-7-bilhoes-e-incremento-na-producao-de-soja

Odilla, F. (2010, November 2). Estrangeiros compram 22 campos de futebol por hora [Foreigners buy 22 soccer fields per hour]. *A Folha de S. Paulo.* Retrieved from http://www1.folha.uol.com.br/fsp/poder/po0211201002.htm

Oliveira, A. (2010). A questão da aquisição de terras por estrangeiros no Brasil: Um retorno aos dossiês [The question of land purchase by foreigners in Brazil: A return to the files]. *Agrária, 12*(1), 3–113.

Oliveira, G. d. L. T. (2013). Land regularization in Brazil and the global land grab. *Development and Change, 44* (2), 261–283.

Oliveira, G. d. L. T. (2015). Chinese and other foreign investments in the Brazilian soybean complex (Working Paper n. 9). Cape Town: BRICS Initiative for Critical Agrarian Studies.

Oliveira, G. d. L. T. (2016). The geopolitics of Brazilian soybeans. *The Journal of Peasant Studies, 43*(2), 348–372.

Oliveira, G. d. L. T., & Hecht, S. (2016). Sacred groves, sacrifice zones and soy production: Globalization, intensification and neo-nature in South America. *The Journal of Peasant Studies, 43*(2), 251–285.

Oliveira, G. d. L. T., & Schneider, M. (2016). The politics of flexing soybeans: China, Brazil, and global agroindustrial restructuring. *The Journal of Peasant Studies, 43*(1), 167–194.

Oliveira, H. (2010). Brasil e China: uma nova aliança não escrita? [Brazil and China: A new unwritten alliance?]. *Revista Brasileira de Política Internacional, 53*(2), 88–105.

Patel, R. (2013). The long green revolution. *Journal of Peasant Studies, 40*(1), 1–63.

Perrone, N. (2013). Restrictions to foreign acquisitions of agricultural land in Argentina and Brazil. *Globalizations, 10*(1), 205–209.

Pombo, B. (2012, March 9). Parecer da AGU sobre compras de terras por estrangeiros é criticado [Legal opinion by the Attorney General Office on farmland purchases by foreigners is criticized]. *Valor Econômico.* Retrieved from http://www.valor.com.br/brasil/2563464/parecer-da-agu-sobre-compras-de-terras-por-estrangeiros-e-criticado

Portal, A. Z. (2010). Lula quer por fim a farra de compra de terras no Piauí por estrangeiros [President Lula wants to end the party of foreign acquisition of farmland in Piauí state]. Retrieved from http://www.portalaz.com.br/noticia/brasilia/175349_lula_quer_por_fim_a_farra_de_compra_de_terras_no_piaui_por_estrangeiros.html

Powell, D. (2011, October 10). The dragon's appetite for soy stokes Brazilian protectionism. *Financial Times.* Retrieved from http://blogs.ft.com/beyond-brics/2011/10/10/the-dragon%E2%80%99s-appetite-for-soy-stokes-brazilian-protectionism/

Pretto, J. M. (2009). *Imóveis rurais sob propriedade de estrangeiros no Brasil* [Rural real estate under control of foreigners in Brazil]. Report of the technical cooperation project 'Support for policies and social participation in rural development.' Brasília: PCT-IICA/NEAD (Center for Agrarian Studies and Rural Development).

Qatar News Agency. (2010, October 5). *Saudis to invest $500 million in Brazilian agriculture.* Retrieved from http://farmlandgrab.org/15995

Rasmussen, R., Chow, J.-Y., Nelson, D., Hendricks, E., & Sarvanti, P. (2011). Chinese investments in South American agribusiness: Overview of an ongoing expanding process (Rabobank Industry Note 276). Utrecht: Rabobank International.

RENAI (National Network of Information about Investments). (2009–2015). *Anúncios de projetos de Investimentos* [Investment project announcements]. Multiple issues. Brasília: Ministry of Development, Industry, and Foreign Trade.

RENAI (National Network of Information about Investments). (2012). *Anúncios de investimentos Chineses no Brasil: 2003–2011* [Announcements of Chinese investments in Brazil: 2003–2011]. Brasília: Ministry of Development, Industry, and Foreign Trade.

Reuters. (2009, November 23). Brasil recusa proposta do Irã de compra de terras [Brazil rejects Iranian proposal to purchase farmland]. Retrieved from http://farmlandgrab.org/post/view/9198-brasil-recusa-proposta-do-ira-de-compra-de-terras

Roldão, A. (2007, May 8). Incra ascende sinal de alerta contra compra de terras por estrangeiros [INCRA lights warning sign about purchase of farmland by foreigners]. *O Estado de São Paulo.* Retrieved from http://politica.estadao.com.br/noticias/geral,incra-acende-sinal-de-alerta-contra-compra-de-terras-por-estrangeiros,29887

Roldão, A. (2008a, March 5). Para Incra, compra de terra na fronteira é proibida pelas leis do País [Purchase of land along the border is prohibited by the country's laws, according to INCRA.] *O Estado de São Paulo,* p. A8.

Roldão, A. (2008b, March 6). Venda de terras para estrangeiros cresce sem controle, afirma Incra [Sale of land to foreigners increases without control, affirms INCRA]. *O Estado de São Paulo*. Retrieved from http://politica.estadao.com.br/noticias/geral,venda-de-terras-para-estrangeiros-cresce-sem-controle-afirma-incra,135638

Salidjanova, N. (2011, March 30). *Going out: An overview of China's outward foreign direct investment* (Staff research report). Washington, DC: US-China Economic & Security Review Commission.

Sant'anna, L. (2014, January 2). Chineses desistem de plantar e agora financiam e exportam soja brasileira [The Chinese give up farming and now finance and export Brazilian soy.] *O Estado de São Paulo*, p. B1.

Sauer, S. (2010). Demanda mundial por terras: 'land grabbing' ou oportunidade de negócios no Brasil? [Global demand for land: 'land grabbing' or business opportunity in Brazil?] *Revista de Estudos e Pesquisas sobre as Américas* (*Journal of Studies and Research on the Americas*), 4(1), 72–88.

Scoton, L. E., & Trentini, F. (2011). *A limitação à aquisição de propriedades rurais por pessoas jurídicas de capital estrangeiro: grupos de interesse e efeitos socioeconômicos* [The limitation on the acquisition of rural property by companies controlled by foreing capital: interest groups and socioeconomic effects]. Annals of the first academic debate: Conference on development, pp. 1–19. Brasília: IPEA.

Shree Renuka. (n.d.). *Shree Renuka Sugars Ltd.: Brazil*. Retrieved from http://www.renukasugars.com/en/renuka-do-brazil-S-A.html and http://www.renukasugars.com/en/renuka-vale-do-ivai-S-A.html

Silva, M., & Costa, L. (2012). A indústria de defensivos agrícolas [The pesticide and herbicide industry]. *BNDES Setorial*, 35, 233–276.

Smaller, C., Wei, Q., & Liu, Y. (2012). *Farmland and water: China invests abroad*. Winnipeg: International Institute for Sustainable Development.

Soy and Corn Advisor. (2010, April 29). *China buying land in 40 countries to produce food destined for China*. Retrieved from http://www.soybeansandcorn.com/news/Apr29_10-China-Buying-Land-in-40-Countries-to-Produce-Food-Destined-For-China

Staufer, C. (2014, April 4). Big Chinese soy project in Brazil: so far, just an empty field. *Reuters/Chicago Tribune*. Retrieved from www.chicagotribune.com/business/sns-rt-us-brazil-china-soybeans-20140404,0,929120.story

Stratfor. (2008, 30 April). *China: 'Going out' for food security*. Retrieved from https://www.stratfor.com/analysis/china-going-outward-food-security

The Economist. (2009, May 21). Buying farmland abroad: Outsourcing's third wave. Retrieved from http://www.economist.com/node/13692889

The Peninsula. (2010, August 29). Hassad Food to make own brand of rice in Qatar. Retrieved from http://farmlandgrab.org/post/view/15115

Todeschini, M., & Rydlewski, C. (2012). A invasão Chinesa [The Chinese invasion]. *Exame: Negócios*, 6(62), 70–85.

UKTI (UK Trade & Investment). (2014). *Chinese outward investment: Keeping up the pace*. Retrieved from https://www.gov.uk/government/publications/chinese-outward-investment-keeping-up-the-pace/chinese-outward-investment-keeping-up-the-pace

Valor Econômico. (2009, December 29). Estrangeiros aceleram aportes no campo [Foreigners increase investments in the countryside]. Retrieved from http://www.valor.com.br/arquivo/801237/estrangeiros-aceleram-aportes-no-campo

Valor Econômico. (2010, April 27). Chineses querem mais terras no Brasil [The Chinese want more land in Brazil]. Retrieved from http://farmlandgrab.org/post/view/15504

Vital, N. (2010, November 11). Chineses fecham acordo de US$7.5 bilhoes com Goiás [The Chinese seal a deal worth 7.5 billion USD with Goiás state]. *Exame*. Retrieved from http://exame.abril.com.br/blogs/aqui-no-brasil/2010/11/11/chineses-fecham-acordo-de-us-75-bilhoes-com-goias-3/

von Braun, J., & Meinzen-Dick, R. (2009). *'Land grabbing' by foreign investors in developing countries: Risks and opportunities* (IFPRI Policy Brief 13). Washington, DC: International Food Policy Research Institute.

Wesz Jr. V. (2016). Strategies and hybrid dynamics of soy transnational companies in the southern cone. *The Journal of Peasant Studies*, 43(2), 286–312.

Wilkinson, J., Reydon, B., & Di Sabbato, A. (2012). Concentration and foreign ownership of land in Brazil in the context of global land grabbing. *Canadian Journal of Development Studies/Revue Canadienne D'études du Développement*, 33(4), 417–438.

Wilkinson, J., & Wesz Jr. V. (2013). Underlying issues in the emergence of China and Brazil as major global players in the new south-south trade and investment axis. *International Journal of Technology Management & Sustainable Development, 12*(3), 245–260.

World Bank. (2010). *Rising global interest in farmland: Can it yield sustainable and equitable benefits?* Washington, DC: Author.

Zhou H., Liu Y., & Guo J. (2011). 外商投资发展中国家土地的分析及对我国的启示 [Analysis of FDI in land developing countries and its implications]. *Zhongguo Ruankexue* (*China Soft Science*), *9*, 41–54.

Zoomers, A. (2010). Globalisation and the foreignisation of space: Seven processes driving the current global land grab. *The Journal of Peasant Studies, 37*(2), 429–447.

Zou J., Long H., & Hu Z. (2010). 国际土地资源开发利用战略初探 [A preliminary study of strategies for China's participation in the exploitation of land resources abroad]. *Ziyuan Kexue* (*Resource Science*), *32*, 1006–1013.

Land control and crop booms *inside* China: implications for how we think about the global land rush

Saturnino M. Borras Jr., Juan Liu, Zhen Hu, Hua Li, Chunyu Wang, Yunan Xu ⓘ, Jennifer C. Franco and Jingzhong Ye

ABSTRACT

This paper aims to broaden the scope of analysis of the contemporary global land rush by examining crop booms not only outside, but inside China; and investment flows not only from China, but also within and into China. It does so by examining the eucalyptus and sugarcane sectors in southern China, which have witnessed investment booms during the past decade, with capital being infused by both domestic capital and foreign capital, including Finnish, Indonesian, and Thai companies. Our argument addresses three key issues: (a) explaining why foreign and domestic companies enter into a multitude of lease and grower contracts involving holders of micro-plots, (b) revisiting the notion of extra-economic coercion, and (c) a critique of thinking about flows of large-scale investments centred primarily on nationality. These issues are central in current debates in the land grabs literature, and our study offers a different perspective from dominant narratives.

Introduction: the global land rush and crop booms inside China

While the role of national governments and domestic capital has been appropriately situated in most analyses of the global land rush (Borras & Franco, 2012; Edelman, Oya, & Borras, 2013; Wolford, Borras, Hall, Scoones, & White, 2013), the role of foreign investors remains an important point of controversy and interest. The role played by the BRICS countries as important sources of cross-border investment flows as a key context in current global agrarian transformations has been underscored (Scoones, Amanor, Favareto, & Qi, 2016). China is always implicated in crop booms – principally as a direct source of investments, or as a key part of the wider context (Bräutigam & Zhang, 2013; Hall, 2011). Crop booms include oil palm, sugarcane, maize, soya, industrial trees, and rubber. In turn, these booms are a result of increased market demands, partly in response to multiple crises in the capitalist world: food, energy/fuel, climate change, financial, leading to the rise of 'flex crops and commodities' (Borras, Franco, Isakson, Levidow, & Vervest, 2016; Kröger, 2016). Crop booms have spurred a similar 'boom' in academic studies of this phenomenon. Most of these studies take the BRICS countries and middle-income countries (MICs) as important contexts. It is impossible to talk about soya in southern America without talking about Chinese demand

This article was originally published with errors. This version has been corrected. Please see Corrigendum (http://dx.doi.org/10.1080/14747731.2018.1417210).

for soya products (see, e.g. Oliveira & Hecht, 2016; Yan, Chen, & Ku, 2016); or soya in Bolivia without talking about the soya complex controlled by the Brazilian capital (McKay & Colque, 2016).

The BRICS countries and several MICs that are involved in crop booms experience internal booms themselves: soya is a boom crop both inside and outside Brazil (Oliveira & Hecht, 2016), in all cases involving Brazilian capital; similarly, oil palm is a boom crop both inside and outside Malaysia, in all cases involving Malaysian capital. There are now emerging studies of the complex political economy of 'inside–outside' crop booms (e.g. Brazil–Bolivia, Brazil–Paraguay) that are yielding fresh insights on agrarian transformations and possible future trajectories for the agrarian societies and communities implicated in these.

Yet, we argue that this perspective does not represent a full picture of China's location in the global crop boom, and so in the global land rush. It is rare to see studies that look into the recent crop boom *inside* China (but see Siciliano, 2014 for an exception). During the past decade, at least two agricultural sectors have seen marked expansions of domestic production: sugarcane and industrial trees (eucalyptus and pine). These are concentrated in the southern part of the country, and more specifically in Guangxi Zhuang Autonomous Region (hereafter Guangxi). Our paper aims to contribute to broadening the conversation by examining crop booms not outside, but *inside*, China; and investment flows not from China, but *within* and *into* China. Specifically, we examine crop and commodity booms involving fast-growing trees (eucalyptus) and sugarcane in southern China. These are sectors that have witnessed investment booms with an infusion of capital not only from domestic sources, but also from foreign capital, including Finnish, Indonesian, and Thai companies. We examine this agrarian transformation inside China not in isolation from, but in relation to, the changing context at the international agrarian front.

Three key findings in our paper resonate strongly with the international literature – but in a way that questions dominant narratives in the literature and public debates, as follows: first, the lands acquired in southern China are generally small, even micro-scale, plots. Thus, 'large-scale' land acquisition should not be misconstrued as always involving one large unit of land entered in a land deal; rather scale has something to do with both land and capital. Our study shows that foreign and domestic corporations have acquired and invested in land in southern China despite having to deal with scattered tiny plots requiring a multitude of individual lease and grower contracts. At a glance, this goes against textbook arguments from new institutional economics about the need to avoid high transaction costs and risks, yet foreign and domestic companies have managed to gain control of several hundred thousand hectares of land through such a process. Topographically, these sugarcane and eucalyptus farms look like an endless, massive quilt, made of tiny patches sewn together. The article traces and examines the political and institutional bases of the crop boom in this unlikely institutional setting. Indeed, this case validates the working definition of land grabbing as a form of 'control grabbing', the notion suggested by Borras, Franco, Gomez, Kay, and Spoor (2012), in which ownership of land is less important than effective control of its use.

Second, and closely related to the first, the notion of 'extra-economic coercion' has been central to the land grabs debate, and some contributions have offered nuanced discussion of this issue (Hall, 2013; Scoones, Hall, Borras, White, & Wolford, 2013). But at times public debates tend to implicitly equate extra-economic coercion with force, violence and intimidation. Our study shows that coercion can take much more subtle forms, such as a combination of price incentive/disincentives, land-use zoning, and production quotas. Taken together and when enforced, villagers do not have to be physically and forcefully asked to give up land or particular land uses to be subject to extra-economic coercion.

Third, the nationality of companies and how this plays out in land deals is an important issue, but much of the literature tends to portray only a one-way trajectory: China as the origin of diverse investments in large-scale land deals. Our study examines two other trajectories: Chinese investors investing within China, and foreign companies investing inside China – both requiring a significant degree of land acquisition and control. This reminds us that the fundamental logic of capital is that it will go wherever it can generate profits, regardless of nationality and national borders. This insight shares something in common with the findings by van der Ploeg, Franco, and Borras (2015) in their study of land concentration and land grabbing inside the European Union. This also ties in neatly with Schneider's analysis of Chinese agribusiness companies expanding their clout both outside and inside China (Schneider, 2017), and situates recent scholarships on agrarian capitalism in China within current debates around the global land rush and crop booms (Yan & Chen, 2015; Ye, 2015; Zhang, Oya, & Ye, 2015).

The remainder of this paper is organized as follows. The next section provides an overview of the two crop booms: eucalyptus and sugarcane. It includes a discussion of how foreign capital is heavily involved in relevant sectors inside China. The third section analyses how specific institutional conditions related to land and labour, and subtle forms of 'extra-economic coercion' (land-use and crop/commodity zoning and quotas for crop/commodity production) have facilitated crop and commodity booms. The fourth part examines relations and interactions among key actors, including cleavages and tensions among relevant state and non-state actors. The fifth section focuses on these tensions and emerging limits to the further expansion of these boom crops. The paper ends with some insights that will hopefully enrich broader discussions of the global land rush and the agrarian transformations associated with it.

Crop and commodity booms inside China

Eucalyptus

The demand in China for wood forest products has markedly increased in recent years. China has become the world's second largest wood consumer, with consumption of 4990 million m^3 in 2012 (Qin & Yu, 2014). This demand could not be fully met from pre-existing domestic production. Thus, China has also become the world's largest importer of wood and wood products (FAO, 2013; see also Figures 1 and 2). Yet, the increase in volumes imported could not by itself meet the

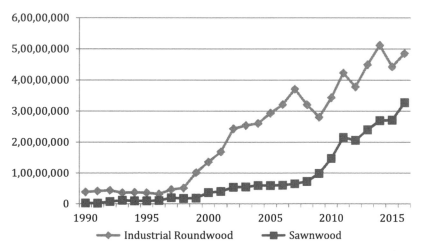

Figure 1. Quantity of industrial roundwood and sawn wood imported by China, 1990–2016 (m^3). Source: FAOSTAT.

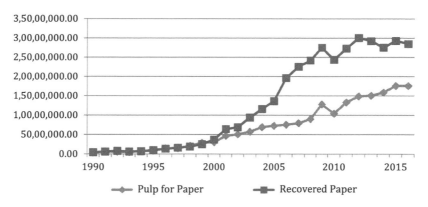

Figure 2. Quantity of imported pulp for paper and recovered paper imported by China, 1990–2016 (tonnes). Source: FAOSTAT.

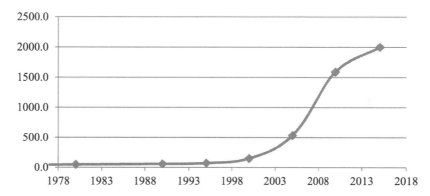

Figure 3. Eucalyptus production in Guangxi, China, 1977–2015 (1000 ha). Sources: Li (2008, p. 40) for data for 1977–2005; Guangxi Forestry Department (2010) for the 2010 entry; Qin and Yu (2014) for the 2014 data.

marked rise in demand, and the expansion of domestic production through the promotion of fast-growing tree plantations has thus become a compelling complementary approach.

Eucalyptus has become the favoured species in new tree plantations, and Guangxi (the star province) within China. To date, 2 million ha of land in Guangxi are planted with eucalyptus, accounting for 21% of China's annual wood supply in 2014 (Qin & Yu, 2014) (see Figure 3).

Sugarcane

Sugar consumption in China has expanded over the years, and the country is currently a net sugar importer (see Figure 4 for importation trends over the past four decades). However, during the same period, domestic sugar production expanded significantly. Sugarcane accounts for about 90% of sugar production in China (USDA GAIN, 2016). Guangxi, Guangdong, and Hainan are traditionally the areas within China where sugarcane has been grown, and have become the most important production zones in the present. Guangxi accounted for more than 60% of the national cultivated area, and thus cane production, in 2015 (see Figure 5), and has seen a great increase in the area planted over the past 15 years (but also a small recent decline).

The 13th five-year plan (2016–2020) of the Chinese government implies that government's intention is to gradually reduce imports and boost domestic sugar production. Both the central

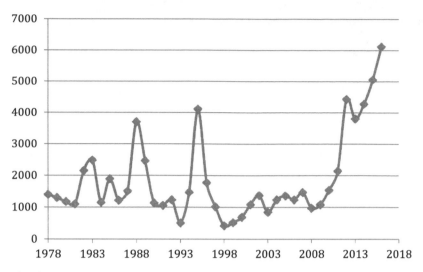

Figure 4. Centrifugal sugar imports to China, 1978–2016 (1000MT). Source: USDA.

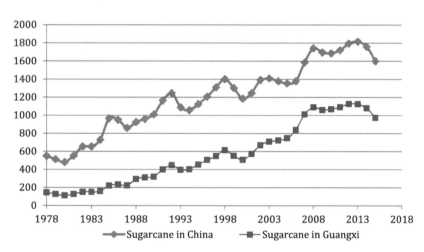

Figure 5. Area planted to sugarcane in China and in Guangxi, 1978–2015 (1000 ha). Source: *China statistics yearbook* (2016).

government and the Guangxi government have begun to provide subsidies and financial support to farmers to enable them to increase yields and reverse the decline in sugarcane acreage, which is partly due to rising production costs. For instance, the Guangxi government has already started providing sugarcane farmers with subsidies of RMB 400 Yuan per mu for seeds, farm machinery, mulching film, and fertilizer. The goal is to put 5 million mu (333,300 ha) of 'double high' sugar cane (high in relation to both 'sugar content' and 'yield') into production (USDA GAIN, 2017).

Foreign and domestic companies in crop and commodity booms inside China

Eucalyptus

Before the 2000s, foreign direct investment (FDI) accounted for a small proportion of investments in the forestry sector, with a cumulative amount of US$1.1 billion before 2000 (Liu, 2002). In 2002–

2012, the cumulative amount of FDI was US$6.24 billion (*China forestry statistical yearbook*, 2003–2013), although it comprised just 1% of the total investments in the forestry sector. Around 2001–2002, the government encouraged further vertical integration in the production and processing of wood and wood products. Companies engaged in vertically integrated wood processing, pulp and paper production have become major players in the sector. Table 1 profiles the leading companies in the pulp and paper-making sector in China.

The Indonesian company Asia Pulp and Paper (APP) first entered China in 1992[1] and established joint ventures with Chinese paper and pulp companies. In 1994, APP established its first direct plantation base in Guangdong Province, which initiated APP's development strategy of integrating forest–pulp–paper in China. Since 1996, APP has successively invested in wholly owned subsidiaries in eastern and southern parts of China and set up its headquarters in Shanghai. As stated by a manager of APP, 'a joint venture acts as a learning tool during the early stage of corporate expansion; however, wholly owned subsidiaries are preferred once ways of communication have been established' (APP-China, 2011). After more than two decades of operations, APP's asset base in China has reached a value of US$22 billion through ownership of over 20 pulp and paper enterprises and more than 20 plantations, covering more than 300,000 ha on the Yangtze River and Pearl River deltas. APP's main business categories in China include copperplate paper, pulp, industrial paper, packaging paper, household paper, and carbonless copy paper. China's 'forest–pulp–paper integration' strategy not only saved APP from the financial and debt crises in the early 2000s, but also helped the company to consolidate its position in the Chinese market and its control over much forest land in China. It has to be noted that land for industrial tree plantations comes from two land categories: state forest farmland and collective forest land.

Stora Enso's operation started in 2002, and now involves 90,000 ha and an integrated board and pulp mill (under construction at the time of our research). The Guangxi investment is the largest ever in the history of Stora Enso. It was planned that the mill's board machine, with a capacity of 450,000 tonnes per year, would be operational by early in 2016, and later a chemical pulp mill would be set up. The eucalyptus plantations will provide the raw materials required for processing. The International Finance Corporation (IFC) has a 5% share in the project, Stora Enso owns 83%, and

Table 1. Major pulp and paper-making enterprises in China, as of 2017.

Company	Origin	Plantation sites	Size of direct plantation operation
Shandong Chenming Paper Group Co., Ltd	Chinese	Hubei, Jiangxi, Hunan, Guangdong, Guangxi	Not clear
Shandong Sun Paper Industry Joint Stock Co., Ltd	Chinese/joint venture with International Paper	Laos and Vietnam (no information of their operation inside China)	100,000 ha
Nine Dragons Paper (Holdings) Limited	Chinese	Vietnam; no details inside China	Not clear
China Paper Corporation	Chinese	Hunan, Guangdong	Not clear
Lee & Man Paper Manufacturing Co., Ltd	Hong Kong	Guangxi, Chongqing, etc.	More or less 3400 ha in Chongqing
APP	Indonesian	Hainan, Guangdong, Guangxi, Yunnan, and others	300,000 ha
Stora Enso	Finnish/Swedish/IFC/GX Forest Group	Guangxi	90,000 ha
Oji Holdings Corporation	Japanese	Guangxi	7500 ha
International Paper	US	Invested in manufacturing; no plantation operation	–
UPM	Finnish	Invested in manufacturing; no plantation operation	–

Source: Authors' collated data from company websites.

13% is owned by Guangxi Forestry Group Co. Ltd and Beihai Forestry Investment & Development Company Ltd.

There are also Japanese and Hong Kong investors in the sector, while two other leading companies in the world – International Paper (US) and UPM (Finnish) – have invested in manufacturing plants. When large Chinese companies joined up with these foreign companies, the result was a marked boom of the sector, a spike graphically illustrated in Table 1. As latecomers to the sector, many Chinese companies could not gain access to good forest land inside China, and these companies began to leave China in search of a resource base, especially in Southeast Asia. For example, Shandong Sun Paper is planning to establish 100,000 ha of eucalyptus plantations in Savannakhet province in central Laos (World Rainforest Movement, 2009).

These foreign and domestic companies deploy different strategies to gain access to forest land in China.[2] First, *companies lease land* from local communities and state forest farms for period that range up to 30 years, and then establish and manage the plantations themselves. Companies typically assume full responsibility for financing projects and make annual lease payments to the community and state farms, but manage the site directly. Stora Enso favours this strategy. Second, *companies establish* a *joint financing arrangement* between the company and a private investor, in which they share the cost of plantation development and the latter assumes responsibility for managing the site, harvesting the wood, and delivering it to the mill site. Typically, the company receives a pre-defined portion of the harvest, while the private investor retains the right to sell the remainder to the mill at the prevailing market price. Some of APP's plantations are operated in this model. Third, a *production-sharing arrangement is established* between the company and the local community or state-owned forests, as land-owner. The company assumes full responsibility for financing the plantation development on community land, and the community is responsible for managing the site. At the end of the rotation, the wood harvested is divided between the company and the community according to an agreed ratio. Often the company provides a guarantee that it will purchase the community's portion of the wood at a pre-determined price. Oji company operate their plantations using this model. Fourth, *companies secure wood procurement contracts with the region's state-owned tree farms or rural communities* which have substantial existing plantation areas. For the collective land, APP invests in the inputs required for afforestation and tending trees, as well as in local infrastructures, while farmers and the community manage it by themselves. This is an arrangement similar to contract-farming and releases industrial companies from responsibility for the management work of the plantations. For elaboration of these strategies, see Xu (in press-a).

Sugarcane

More than 30 sugar companies have now established their mills in Guangxi, among which five companies are state-owned or state-controlled, five are controlled by foreign companies, and the rest are domestic private enterprises. For the sugar industry, there is a zoning rule for each mill, which means that sugar companies cannot expand their sugarcane area through land acquisition in areas outside their allocated zone. It is partly for this reason that several Chinese companies have begun to expand their sugarcane base by setting up operations in Vietnam and Myanmar to increase the supply of raw materials for sugar milling (see Borras & Franco, in press). This form of expansion has been encouraged and supported by local government bodies in China, since tax from sugar milling has become the main income source for local government in these areas.

One of the largest players in the sector is the Guangxi Nanning East Asia Sugar Group (EASG), established by a parent company, the Thai Mitr Phol Sugar Group in October 1993. At present,

EASG operates seven sugar plants in Chongzuo City as joint ventures, with most of the sugar plants controlled in the past by the local government bodies. AB Sugar China's involvement in the Chinese sugar industry began in 1995 with a joint venture in a mill in Guangxi, known as BoQing. Collectively, their five sugarcane factories can process 5 million tonnes of sugarcane and produce over 600,000 tonnes of sugar annually. AB Sugar China has a policy of 'comprehensive utilization' for all co-products; so the mills also produce and sell bagasse, molasses, fertilizer and alcohol. Yunnan Power Biological Products Group has 14 companies in China, Laos and Myanmar. The factories and mills owned by this company process sugarcane, cassava and natural rubber, with sugar, alcohol, and rubber as their main products.

Land and labour as dimensions of crop booms

With reference to Tania Li's framing of the land and labour dimensions of global land grabbing (Li, 2011), the case of Guangxi demonstrates a diverse mix of land-labour relations in the context of land investments. There are cases where companies are in need of land but not labour, but others where both land and labour are needed by companies. The crop/commodity boom inside China has been accompanied by a widespread conflict among villagers, local government, land and labour brokers, and companies (Li & Wang, 2014). However, many of the contentious issues are not about displacement or dispossession, as in many cases elsewhere in the world (Hall et al., 2015). Rather, much of the tension centres around the terms and conditions under which villagers are incorporated into these emerging enterprises within the crop/commodity boom. The structural and institutional conditions in the crop boom sites thus shape the political character of tensions over land and labour.

A large-scale rural–urban migration process has been taking place in China since the early 1990s (Ye, Wu, Rao, Ding, & Zhang, 2016). About 20 million labourers left rural communities to work in urban areas in 1993. This figure rose to more than 22 million in 1998 (Rural Social and Economic Survey Office, 2005), and pole-vaulted to 220 million in 2008 and 260 million in 2013, respectively (National Bureau of Statistics of China, 2014). Geographically speaking, these large-scale outflows of workers are concentrated in six provinces and municipalities: Guangdong, Shanghai, Beijing, Jiangsu, Tianjin and Fujian. In Guangxi and eight other provinces, the outflows of population are higher than the inflows (Lu & Zhou, 2014). National population census surveys of 2000, 2005 and 2010 revealed that the net outflows in these nine provinces amounted to 2.55, 2.65 and 4.57 million people, respectively, in these years. In Guangxi, the proportion of the population engaged in agricultural production decreased from 70% in 1980 to 28% in 2012. These shifts have set the stage for subsequent crop booms in Guangxi (Guangxi Employment Department, 2014).

Official statistics (Guangxi Employment Department, 2014) show that the number of 'peasant workers' in Guangxi reached 11.65 million in total in 2013, of which 8.98 million migrated outside Guangxi to work. Approximately 6.23 million (or 69.4% of the out-migrants) went to Guangdong province with the majority of those (60.9%) working in the manufacturing sector Thus, seasonal and cyclical migration had already been prevalent by the time that foreign and domestic capital moved into the agricultural sector in Guangxi in the 1990s–2000s.[3]

Labour dynamics have major implications for the politics of rural land. There are two types of ownership of rural land in China: 'state-owned' and 'collective owned'. Most of the state-owned land was originally under the management of state farms and state forest farms, while collective land usually refers to land owned by the village, or sub-village entities such as 'production teams'. From the late 1970s until 1983, China implemented the Household Responsibility System (HRS) for the distribution of collective arable land, wherein each household took responsibility for a plot of land and its associated entitlements and responsibilities (Kung & Liu, 1997).

Land distribution in a village was usually conducted by categorizing all village arable land into different types according to fertility and productivity, and then equitably distributing these to households. What this means is that one household may hold multiple small plots dispersed in different locations within the village, each with different physical conditions, for example, in relation to soil fertility and access to irrigation. In principle, every household in a village should have some share of both flat irrigated land, and rocky and hilly land. This pattern of land distribution, originally borne out of a concern for egalitarianism, would play a key role in facilitating the subsequent crop boom. In 1993, the government declared that farmers' land tenure rights should be guaranteed in terms of 30-year contracts, hoping that peasants would be encouraged to invest in their land.[4]

Similar reforms were conducted in relation to the collective forest tenure system in the early 1980s, in order to motivate farmers to invest in tree planting and forest management. Slightly different from the arable land reform, the forest land reform was initially conducted by allocating small parts of collective-owned forest land to individual households as 'private mountains', and contracting the majority of collective forest land to these households as 'responsibility mountains'. By 1986, more than 70% of all collective-owned forest land had been allocated and contracted to farmer households, for an even longer period than arable farming land (Yin, Xu, & Li, 2003). Moreover, another type of rural land is known as the 'Four Wastelands' (*Sihuang*), including 'waste mountains', gullies, hills and riverbanks. In many provinces, wasteland was not clearly allocated to households in the beginnings of the reform, even if some wasteland was contracted to peasants at low or zero cost. From the late 1980s, however, some provinces decided to address the problem of soil erosion on barren, hilly lands. In this context, wasteland use rights were put up for open auction (Zhang, Liu, & Wang, 2002a, 2002b).

When the sugar industry started to expand in Guangxi in the late 1980s, people were encouraged to develop and reclaim wasteland in order to grow sugarcane. By the 1990s, this initiative had become very popular among villagers, who started to reclaim wasteland on a large scale. Auctions for waste mountains and waste hilly land accelerated afforestation and facilitated the expansion of fast-growing trees and plantations in Guangxi, accompanied by the government's *fast-growing, high yield* programme, which is meant to convert fragile farmland to forestland. This policy has been invoked at times to enable the conversion of sugarcane land to eucalyptus plantations. However, villagers tend to discover the full extent of the 'new' and 'real' value of so-called wasteland only after foreign companies (specifically Stora Enso and APP) have begun to lease land for eucalyptus production, from both state forest farms and collectives (Xu, in press-c).

State policies and the crop booms

Government policies have facilitated crop booms in China. These policies have resulted in the two sectors, eucalyptus and sugar, not only co-existing, but also complementing and contradicting one another. Their coexistence is thus marked by both tension and synergy.

Development of the Chinese sugarcane sector has taken place in various stages. In the 1950s, the government tightly regulated the sugar sector. In 1955, the Chinese Ministry of Light Industry decided to rapidly develop sugar processing in Guangxi. Sugar factories were established, and sugarcane was in great demand. In the next year, Guangxi government established a sugarcane pricing regime in which all sugarcane had to be sold to state-owned sugar factories at fixed prices.[5] In order to safeguard the adequate supply of raw material for sugar factories, government encouraged the villagers to plant sugarcane by providing additional incentives: (a) in 1972, a policy was introduced that villagers engaged in sugarcane production in certain government-determined sugar zones should be provided with rationed foodstuffs; (b) sugarcane farmers were given access to loans and fertilizer; and (c) investments were made in infrastructure in sugarcane producing regions, especially in water conservation and road construction.

In the 1980s, the HRS allowed households to manage agricultural production on their own, while farmland remained in the ownership of the rural collective. With the initial establishment of a market economy in China, some crops were freely traded in market. Until 1992, the price of *sugar* was thus market-based, while the price of *sugarcane* was determined in government-led negotiations between sugarcane farmers and factories. In 2001, China became a member of the World Trade Organization (WTO). One of China's commitments within the WTO was that the minimum import quota for sugar in 2002 would be 800,000 tonnes, which would increase 1.5% annually, and the tariff would be no more than 15% after 2005. With changes in patterns of consumption and growing demands for sugar, there has been a marked rise in the importation of sugar, which is cheaper than domestically produced sugar. To protect China's domestic sugarcane sector, government passed the Interim Measures for Sugar (*Tangliao Guanli Zanxing Banfa*) in 2002, which remains the most important form of regulation of the sugarcane sector. It also established the Sugarcane Region Regime (SRR), which has been in force since then.

Land-use zoning has become a central pillar of current sugarcane policy. The aim was to bind sugarcane production by villagers together with sugar factories' interests. Generally speaking, a county is divided into sugarcane regions, each connected to a sugar factory. In these regions, a sugar factory enters into contracts only with the sugarcane farmers who plant sugarcane for that specific factory. A sugarcane farmer cannot sell sugarcane to non-contract factories, even if the latter offers a higher price. Penalties are imposed for violations of these rules. For example, in 2006–2007, six sugar companies were penalized for buying sugarcane from other sugarcane regions.

Every year before the sugarcane harvest, the provincial government price bureau issues a mandatory price for sugarcane. In the crushing season, the price bureau issues a second sugarcane price, which is calculated according to the price of sugar. If the second price is higher than the first, the excess should be given to villagers; however, in the reverse, it is not necessary for villagers to return money to factories. Interestingly, all the sugarcane farmers we interviewed in the field said that they had never received any compensation for price differences.

Moreover, planting of eucalyptus trees is not permitted in some sugarcane regions, even on hilly lands, ostensibly because eucalyptus plantations damage soil and water. Article 20 of the Nanning City Drinking Water Source Protection Regulations of 2014 states that industrial eucalyptus should not be planted in and around sources of drinking water. This is one major bone of contention between sugarcane and eucalyptus producers. Local government officials have begun to be quite strict in enforcing the rule. However, it is public knowledge in Guangxi that the main reason why government officials are not happy with the eucalyptus sector is because it does not provide them with the substantial taxes yielded by the sugarcane sector.

The SRR is unquestionably beneficial to central and local government. Since 2006, the villagers have not paid taxes on their sugarcane production. Theoretically speaking, villagers can plant any crop in their lands, but if they plant sugarcane, they do not have to pay taxes, while the sugar companies secure enough raw material for sugar production – and in turn the government can obtain tax income from companies.

Alongside land-use zoning, the sugar quota system has also played a key role in shaping the development of the sector. In the 1960s, central government required Guangxi to produce 1.8 million tonnes of sugarcane and 35,000–40,000 tonnes of sugar. In order to complete the task, Guangxi assigned Chongzuo County and 11 other counties the status of sugarcane production zones. In

1977, Guangxi ruled that the state farm, Jin Guang Farm, and 10 other state-owned farms should make sugarcane their main crop, and that more than 70% of their arable land should be used to plant sugarcane. In 1988, the Chinese state council suggested that Guangxi makes full use of its 8 million mu[6] of dry, hilly lands, with 75% of these lands being suitable for sugarcane plantation.

Currently, the 'high in sugar content and yield' Sugarcane Production Base (HSPB, Shuanggao jidi) is the key focus of the quota system. Guangxi has decided to develop 5 million mu for its HSPB. In 2014, the Guangxi government allocated this task among its various regions, involving a total area of 0.5 million mu. Each layer of the government bureaucracy parcelled out the task to its lower units, until the task was distributed among the villagers. This is called the 'level-to-level contract responsibility system' or a 'sugarcane quota system with Chinese characteristics'.

What this suggests is that the sugarcane sector contributes substantially to the value added to the local economy and to local government taxes. This makes it a much more favoured sector than eucalyptus, and is one reason why there is much tension and conflict among villagers, state farms, companies and local government units, and within and between the sugarcane and eucalyptus sectors.

According to the Chinese tax system, government, especially local government, can only obtain tax from the processing of sugarcane, not from the production of sugarcane.[7] Usually the sugar enterprise should pay enterprise income tax (EIT) and value-added tax, both of which accrue to the central and local governments.[8] Similarly, the farmers pay no tax for their eucalyptus plantation, but, if a eucalyptus processing factory exists, the government, especially local government, can collect tax from it. There are not as many eucalyptus manufacturing plants in Guangxi as there are sugarcane mills; hence, local government officials tend to dislike eucalyptus, despite central government's interest in promoting eucalyptus production.

Local government usually receives more than half of the sum of sugar taxes. In Chongzuo, the 'sugar capital of China', two sugar factories paid a total tax of 45,750,000 Yuan, which accounted for 70.18% of county revenue in 1993 (Guangxi Statistics Bureau, 2004). In 2011, Guangxi Dongmen Nanhua Sugar Company paid local tax of more than 0.1 billion Yuan. According to a statement, the tax paid by the two sugar factories usually accounts for about 50–60% of the total tax paid in Fusui County.

Villagers engaged in sugarcane production appear to be doing so of their own accord, responding to a variety of institutional incentives. However, critical analysis suggests that these incentives constitute a subtle form of coercion. The eucalyptus sector thrives in part because many households do not have sufficient labour that can be mobilized for labour-intensive sugarcane production, when compared to eucalyptus production, which does not require as much labour. Yet incentives are skewed towards sugarcane production.

Key actors and tensions among them

Crop booms in China have generated complex interactions between an array of state and non-state actors, including foreign companies, in political dynamics marked by both tension and synergy. As mentioned earlier, conflict has been widespread in these two sectors in Guangxi, and many revolve around the terms of incorporation of villagers into emerging enterprises, a type of conflict that is also significant elsewhere in the world (Hall et al., 2015; Xu, in press-b).

Sugar: farmers, companies, and local government

There is much observable conflict over control of land among farmers, sugar companies and the state. The main concerns of these actors are in relation to what types of crops should be planted, how they are to be planted, and on which land. They are united in the calculation that there is much potential profit to be realized in these enterprises, but tensions tend to increase when contracted villagers run into financial difficulties.

At the height of the sugarcane boom (and just before the price decline of 2015), villagers, companies and local government officials all appeared to be happy, and popular folk songs depicted a 'sweet' situation for all concerned. In an interview, a company official exclaimed: 'There used to be huts, muddy roads and grasses, now there are paved roads, piped water, and these peasants are able to build up several-story houses, or "sugarcane houses" (*Gan zhe lou*), thanks to sugarcane.'[9] A sugarcane farmer said: 'At that time, the sugar company paid us a lot, and they paid us quickly – only ten days after sending the sugarcane into the mill factory, we got the payment.'[10] Sugarcane zones were dynamized and expanded, and specialized sugarcane farmers emerged, defined by sugar companies as those with a minimum of 70 mu under sugarcane or producing 400 tonnes of production per year. They were issued with a certificate, and showered with extra benefits, such as leisure travel to Vietnam, Yunnan and Beijing. Nanhua Sugar Company and East Asia Sugar Company provided between 50% and 60% (in some years over 65%) of local revenues annually, worth about 700 million Yuan in 2013, in which Nanhua was responsible for 280 million Yuan and East Asia 420 million Yuan.

However, from around 2010 the sugarcane sector began to experience a number of difficult challenges. Firstly, labour costs soared. Payments for casual labourers in the harvesting season increased from 20–30 Yuan per day in 1996–1997 to 50–60 Yuan per day in early 2000. In 2016, the cost of labour was about 60–80 Yuan per day before the Spring Festival (in late January/early February) and 100–130 Yuan per day after the Spring Festival, when the supply of labour dwindles.[11] Few migrant workers from provinces located in inner China (such as Yunnan and Guizhou) are available to harvest sugarcane at this time, since it coincides with their need to work on their own farms back home, and thus harvesting relies heavily on migrant workers from Vietnam. 'Without Vietnamese workers, we wouldn't be able to harvest the sugarcane', a sugar company employee explained.[12] Indeed, in Chongzuo City, shortfalls of 30,000 labourers in the less busy farming season and up to 50,000 labourers in the harvesting and crushing season are common (Wei, 2014).

Secondly, the price stipulated by the central state for sugarcane has progressively declined, amid increasing levels of imports of cheaper sugar. In the 2010–2011 crushing season, farmers were paid 600 Yuan per tonne. The price dropped to 500 Yuan per tonne in 2011–2012, 475 Yuan in 2012–2013, 440 Yuan in 2013–2014 and 400 Yuan in 2014–2015.[13] Farmers in our interviews often complained of 'zero profits' from sugarcane. Some sugarcane farmers said that at 2016 sugarcane prices, the net profit per mu would be 1000 Yuan if a farmer used only family labour, but the net profit would go down to 300–500 Yuan per mu if labour had to be hired in.

Among the reasons for the declining price of sugar globally is the increase in the supply of sugar from a range of sources aside from sugarcane (such as sugar beet), but perhaps more importantly, the substantial expansion of sugarcane production worldwide. Based on FAO Statistics, 19.3 million ha were planted to sugarcane in 2010 globally. By 2014, this had jumped to 27.1 million ha of land.[14] The rise of 'flex sugarcane' (sugarcane used flexibly for sweeteners, ethanol and other commercial and industrial commodities) may have contributed to the popularity of sugarcane, but at the same time may have contributed to the falling price of sugar (see Mckay, Sauer, Richardson, & Herre, 2016).

The global pattern of increasing areas planted to sugarcane is evident in China too. Using FAO statistics, it is estimated that there were 1.1 million ha of land planted to sugarcane in China in 2000, increasing to 1.69 million ha in 2010 and 1.76 million ha in 2014.[15] However, in some areas, there has been a contraction in the area planted to cane. The Nanhua Sugar Company in Chongzuo City in Fusui County, for example, lost a total of 70,000 mu between 2012 and 2014. There has also been a decline in the numbers of specialized sugarcane farmers. There were about 800 such farmers supplying Nanhua Sugar Company before 2013, but this had declined to 426 by 2014.[16] A majority of the remaining specialized farmers have reduced their production of sugarcane, and some have switched to eucalyptus, banana or other fruit crops, adding further to tensions among and between sub-sectors.

Eucalyptus: farmers, state forest farms, local government, and foreign companies

The majority of the land that was previously used for sugarcane by Nanhua Sugar Company has been diverted to eucalyptus. The sugarcane boom began earlier than that in relation to eucalyptus, and the latter has occurred partly at the expense of the former in terms of acreage, for the compelling institutional and political reasons discussed above. Land-use conversion from sugarcane to eucalyptus involved nearly 30,000 mu in 2012–2013 in the Nanhua zone alone.[17] Actors involved in the contests between sugarcane and eucalyptus (such as farmers, state forest farms, local government, forestry bureaus, sugar companies, and forest, pulp, and paper companies) have created competing discourses around eucalyptus, advocating either for or against the crop.

In the sugarcane sector, villagers accuse state farms of either encroaching onto their plots or of outright theft of their plots. On the other hand, state farms that tend to aggregate plots into bigger, scaled up operation also accuse farmers of land theft. A state farm official told us:

> Farmers are cunning. They often steal land from our state farms. [Our state farm] lost over 110,000 mu to local farmers in these years and couldn't take the land back. Suing is not the solution. Local government cares more about maintaining stability and harmony (*Wei wen Wei he*), so the arbitration decision is usually biased in favour of farmers. Even if we win the lawsuit, it is difficult to enforce the judgment. Farmers can easily occupy the land again. They cut a circle in the tree trunk, sow linseed after the tree dies, and then claim the land is theirs. When we take the land back through the court, and plant new seedlings, they may flip the trees with a whip, and a row of seedlings dies in a few minutes. We can do nothing about it.[18]

The conflict around eucalyptus is often triangular: villagers versus state farms versus foreign companies such as Stora Enso and APP. These are often widely reported in the media (Li & Wang, 2014; Xu, in press-b) and are similar to those reported in Cambodia and Ethiopia. This is illustrated in an interview with state farm officials:

> We lost 100,000 mu to Stora Enso in 2007. This foreign company invaded in Guangxi in 2002, promising to build a big forest, pulp and paper company in Beihai Municipality. But more than ten years passed, and nothing has happened yet.[19] Stora Enso took forests from ten state forest farms. Guangxi Forest Corporation was set up to deal with it. Stora Enso gives the land rent to Guangxi Forest Corporation and the latter is supposed to give the rent to the state forestry farms. But when this corporation no longer reported to the State Forestry Administration and became a part of SASAC (State-owned Assets Supervision and Administration Commission) in 2013, it refused to give the rent to us. The state forest farms have to rent land from farmers in order to enlarge their operations.[20]

These examples illustrate how industrial crop booms in southern China are generating a range of conflicts between the key actors involved, both within and across the sugarcane and eucalyptus sub-sectors. They involve variable configurations of land, labour, capital and taxes, and tensions between villagers, the state, and Chinese and foreign capital. The institutional regime involves subtle forms of coercion of villagers. Analyses of the global land rush must be adequate to the complexities of cases such as these.

Conclusion

Why would foreign and domestic companies go to the trouble of dealing with thousands of individual lease and grower contracts in the sugarcane and eucalyptus sectors of southern China, given that both the risks and the transaction costs involved are usually seen as being too high? Crop booms in southern China do not appear to fit with a dominant narrative in the literature on the current global land rush, which suggests that 'large scale' refers to a single chunk of land, of say, 30,000 ha, governed by one contract and usually operated with high levels of mechanization. Based on our study, the plausible answer to this question is twofold: (a) high transaction costs are offset by the economic advantages of being positioned inside China, the world's largest market for wood products and a key market for sugarcane products and (b) the Chinese government is willing to install the necessary (and coercive?) institutional rules to ensure that 'smooth and disciplined' acquisition and consolidation of scattered micro-plots takes place. Companies have achieved the scale of operations they desire by dealing with a multitude of individual villagers, who might each have plots as small as 5 mu to offer.

Why would villagers lease their lands to foreign or domestic companies, or enter into grower contracts with them? One important reason is because many small plots are no longer central to the livelihoods of those who had earlier migrated to the cities in search of wage work (Xu, in press-c). For many villagers, this is an unexpected additional source of income, and since neither eucalyptus nor sugarcane require their full-time labour, leases or contracts allow them to retain their main jobs in the cities. A combination of state-driven institutional incentives and disincentives around land, labour, trade and taxation – separately and together – have formed what can be construed as subtle forms of extra-economic coercion of villagers to enter into eucalyptus and sugarcane production, and then get stuck in it.

Why would China allow foreign companies to gain a foothold inside China and get involved in the politics of land control? The answer is that many of these foreign companies were already in China before the recent crop booms took place. They responded to the earlier campaign by the Chinese government to entice capital into China, and entered the country before the 'going out' Chinese policy was announced in 1999 (Hofman & Ho, 2012; Yan & Chen, 2015). When the global land rush kicked in from 2008 onward, some Chinese companies began to look at land inside China, in order to regain ground from foreign investors. At times, there were attempts by the late coming domestic companies to gain ground in the eucalyptus and sugarcane sectors by competing with the established foreign and domestic companies in terms of land control, causing some tensions between them.

Our study suggests that the dynamics of the global land rush revolve around the political economy of land, labour and capital (Bernstein, 2010): who gets which land, labour and capital, how, how much, why, for what purposes and with what implications? These are the key issues to focus on, rather than other aspects such as the nationality of investors, the cross-border direction of capital, the procedural elements of investment processes, and so on, that, although important in themselves, are in the end somewhat secondary.

Notes

1. Note that in March 2001, during the Asian debt crises, APP defaulted on its debt, most of which was subsequently rescheduled at lower values.

2. The exact figures of acreage and output for these different patterns in Guangxi are not available, but the official data show that in 2014, 656,300 mu land has been circled out and cleared for HSPB ('Guangxi Shuanggao',2015) and by the end of 2015, 992,500 mu, with the government grants reaching 1.798 billion RMB (Ministry of Land and Resource of PRC, 2016).
3. For a relevant background discussion on migration and livelihoods, rural and urban, see van der Ploeg, Ye, and Pan (2014); for the dynamic interplay between land and labour, both rural and urban, see Chuang (2015), Andreas and Zhan (2016), Ho and Spoor (2006), and Ye (2015).
4. This decision was announced by the government's Central Rural Work Committee in October 1993. See CPC Central Committee and the State Council (1993).
5. For instance, in 1972 the price of sugarcane rose from 30.29 to 34.94 Yuan per tonne, but at the same time the price of sugar was unchanged. The production cost of sugar increased, and many sugar factories lost money (see Si, 2004).
6. 1 mu equals to 1/15 ha.
7. For background on the politics of taxation, see Kennedy (2007).
8. The Arrangement Measures for Sugar EIT (Zhitangye Qiye Suodeshui Guanli Banfa), adopted by Guangxi autonomous region in 2009, is one of the most important regulations in the system of sugar taxation.
9. Interview with HLQ, an official in Nanhua Sugar Company, March 6, 2015, Fusui County.
10. Interview with Gan, a village leader, March 7, 2015, Fusui County.
11. Interview with HQZ, a village leader, January 5, 2016, Fusui County.
12. Interview with Huang, an official in Nanhua Sugar Company, March 7, 2015, Fusui County.
13. Interview with XP, an official in Nanhua Sugar Company, March 7, 2015, Fusui County.
14. Calculated from data on the FAOSTAT website.
15. Calculated from FAOSTAT website.
16. Interview with XP, an official in Nanhua Sugar Company, March 9, 2015, Fusui County.
17. Interview with LS, an official in Nanhua Sugar Company, January 9, 2016, Fusui County.
18. Interview with an office director, a technician and an accountant in one of the state farms. March 10, 2015. Fusui County.
19. In fact, Stora Enso's investment in this project began in late 2014.
20. Interview with an office director, a technician and an accountant in one of the state farms in Fusui County, March 10, 2015.

Acknowledgements

We thank the journal's peer reviewers for their critical but constructive and useful reviews that helped improved the quality of this paper.

Disclosure statement

No potential conflict of interest was reported by the authors.

Funding

This work is supported by the Ford Foundation [grant number 0135-1532-0], Basic Research Fund of China Agricultural University [grant numbers 2016RW005, 2017RW002], and the Start-up Grant of Northwest A&F University [grant number 2452015349].

ORCID

Yunan Xu ⓘ http://orcid.org/0000-0002-4520-1207

References

Andreas, J., & Zhan, S. H. (2016). Hukou and land: Market reform and rural displacement in China. *The Journal of Peasant Studies, 43*(4), 798–827.

APP-China. (2011). *APP Zhongguo Huanjing yu Shehui Kechixu Fazhan Baogao* [APP-China Sustainability Report 2010]. Retrieved from http://www.app.com.cn/Upload/201401/1390969553.pdf

Bernstein, H. (2010). *Class dynamics of agrarian change.* Halifax: Fernwood.

Borras, S. M.Jr., & Franco, J. C. (2012). Global land grabbing and trajectories of agrarian change: A preliminary analysis. *Journal of Agrarian Change, 12*(1), 34–59.

Borras, S. M., & Franco, J. C. (in press). Towards agrarian climate justice: The urgent need for land redistribution, recognition and restitution. *Third World Quarterly.*

Borras, S. M.Jr., Franco, J. C., Gomez, S., Kay, C., & Spoor, M. (2012). Land grabbing in Latin America and the Caribbean. *Journal of Peasant Studies, 39*(3–4), 845–872.

Borras, S. M.Jr., Franco, J. C., Isakson, S. R., Levidow, L., & Vervest, P. (2016). The rise of flex crops and commodities: Implications for research. *Journal of Peasant Studies, 43*(1), 93–115.

Bräutigam, D., & Zhang, H. (2013). Green dreams: Myth and reality in China's agricultural investment in Africa. *Third World Quarterly, 34*(9), 1676–1696.

Chuang, J. (2015). Urbanization through dispossession: Survival and stratification in China's new townships. *The Journal of Peasant Studies, 42*(2), 275–294.

The CPC Central Committee and the State Council. (1993). *Zhonggong Zhongyang Guowuyuan Guanyu Dangqian Nongye he Nongcun Jingji Fazhan de Ruogan Zhengce Cuoshi* [CPC central committee and the state council: Policies and measures on agricultural and rural economic development]. Retrieved from http://www.chinalawedu.com/falvfagui/fg22016/12025.shtml

Edelman, M., Oya, C., & Borras, S. M.Jr. (2013). Global land grabs: Historical processes, theoretical and methodological implications and current trajectories. *Third World Quarterly, 34*(9), 1517–1531.

FAO. (2013). *FAO yearbook: Forest products.* Retrieved from http://www.fao.org/3/a-i4746m.pdf

Guangxi Employment Department. (2014). *Guangxi Zaiyue Nongmingong Qinkuang Diaocha Baogao* [Report on rural migrant workers from Guangxi to Guangdong]. *Ren Shi Tian Di, 193*(8), 14–19.

Guangxi Forestry Department. (2010). *Zhongda Zaolin Xiangmu Ji Guihua* [Key afforestation projects and plans]. Retrieved from http://www.gxsti.net.cn/dtxx/gxkj/zzq/622830.shtml

"Guangxi Shuanggao Tangliaozhe Jidi Jianshe Kaiju Linghao [A Good Beginning of HSPB in Guangxi]." 2015. *Guangxi News.* Accessed November 9, 2017. http://news.gxnews.com.cn/staticpages/20150505/newgx554837c8-12723184.shtml

Guangxi Statistical Bureau. (2004). *Guangxi statistical yearbook, 2003.* Beijing: China Statistics Press.

Hall, D. (2013). Primitive accumulation, accumulation by dispossession and the global land grab. *Third World Quarterly, 34*(9), 1582–1604.

Hall, R. (2011). Land grabbing in southern africa: The many faces of the investor rush. *Review of African Political Economy, 38*(128), 193–214.

Hall, R., Edelman, M., Borras, S. M.Jr, Scoones, I., White, B., & Wolford, W. (2015). Resistance, acquiescence or incorporation? An introduction to land grabbing and political reactions 'from below'. *The Journal of Peasant Studies, 42*(3–4), 467–488.

Ho, P., & Spoor, M. (2006). Whose land? The political economy of land titling in transitional economies. *Land Use Policy, 23*(4), 580–587.

Hofman, I., & Ho, P. (2012). China's 'developmental outsourcing': A critical examination of Chinese global 'land grabs' discourse. *The Journal of Peasant Studies, 39*(1), 1–49.

Kennedy, J. J. (2007). From the tax-for-fee reform to the abolition of agricultural taxes: The impact on township governments in north-west China. *The China Quarterly, 189*, 43–59.

Kröger, M. (2016). The political economy of 'flex trees': A preliminary analysis. *The Journal of Peasant Studies, 43*(4), 886–909.

Kung, J. K. S., & Liu, S. Y. (1997). 'Farmers' preferences regarding ownership and land tenure in post-Mao China: Unexpected evidence from eight counties. *The China Journal, 38*, 33–63.

Li, M. (2008). *Guangxi Senlin Ziyuan Dongtai Yanjiu* [Research on the dynamics of forest resources in Guangxi]. Guangxi: Guangxi University Press.

Li, T. M. (2011). Centering labor in the land grab debate. *The Journal of Peasant Studies, 38*(2), 281–298.

Li, P., & Wang, X. B. (2014). *Forest land acquisition by Stora Enso in south China: Status, issues, and recommendations.* Washington, DC: Rights and Resources Initiative.

Liu, P. (2002). *Wo Guo Linye Liyong Waizi Jiegou Fenxi* [Analysis of the structure of foreign fund use in China forestry]. *Forestry Economics, 5*, 43–45.

Lu, Q., & Zhou, H. X. (2014). *Renkou Liudong Shijiaoxia de Chenzhenhua Fenxi* [Urbanization analysis from the perspective of labor migration]. *Macro Economics,* (12), 29–33.

McKay, B., & Colque, G. (2016). Bolivia's soy complex: The development of 'productive exclusion'. *The Journal of Peasant Studies, 43*(2), 583–610.

McKay, B., Sauer, S., Richardson, B., & Herre, R. (2016). The political economy of sugarcane flexing: Initial insights from Brazil, Southern Africa and Cambodia. *Journal of Peasant Studies, 43*(1), 195–223.

Ministry of Land and Resource of PRC. (2016). Guangxi Shuanggao Jidi Tudi Zhengzhi Yiwancheng JinBaiwanmu [The Goal of 1 Million Mu of HSPB Reached in Guangxi]. Accessed November 9, 2017. http://www.mlr.gov.cn/xwdt/dfdt/201602/t20160223_1397078.htm.

National Bureau of Statistics of China. (2014). *2013 Nian Quanguo Nongmingong Jiance Diaocha Baogao* [2013 report on national peasant workers]. Retrieved from http://www.gxtj.gov.cn/tjsj/tjgb/nypc/200804/t20080403_2214.html

National Bureau of Statistics of China. (2016). *China statistics yearbook, 2016.* Beijing: China Statistics Press.

Oliveira, G., & Hecht, S. (2016). Sacred groves, sacrifice zones and soy production: Globalization, intensification and neo-nature in South America. *The Journal of Peasant Studies, 43*(2), 251–285.

Qin, R. S. H., & Yu, Y. Y. (2014). *Guangxi Wei Guojia Mucai Zhanlue Chubei Jianshe Fali* [Guangxi contributes to the national strategic conservation for wood]. Retrieved from http://www.forestry.gov.cn/main/72/content-718253.html

Rural Social and Economic Survey Office, National Bureau of Statistics of China. (2005). *Zhongguo Nongcun Zhuhu Diaocha Nianjian* [Annual China rural household survey]. Beijing: China Statistics Press.

Schneider, M. (2017). Dragon head enterprises and the state of agribusiness in China. *Journal of Agrarian Change, 17*(1), 3–21.

Scoones, I., Amanor, K., Favareto, A., & Qi, G. (2016). A new politics of development cooperation? Chinese and Brazilian engagements in African agriculture. *World Development, 81*(1), 1–12.

Scoones, I., Hall, R., Borras, S. M.Jr., White, B., & Wolford, W. (2013). The politics of evidence: Methodologies for understanding the global land rush. *The Journal of Peasant Studies, 40*(3), 469–483.

Si, W. (2004). *Zhongguo Tangye Fazhan Huigu ji Zhanwang* [China sugar industry development: Review and forecast]. *World Agriculture, 3*, 10–13.

Siciliano, G. (2014). Rural-urban migration and domestic land grabbing in China. *Population, Space and Place, 20*(4), 333–351.

State Forestry Administration of China. (2003–2013). *China forestry statistical yearbook, 2003-2013.* Beijing: China Forestry Press.

USDA GAIN. (2016). *China sugar annual report.* Retrieved from https://www.fas.usda.gov/data/china-sugar-annual-0

USDA GAIN. (2017). *China sugar annual report.* Retrieved from https://www.fas.usda.gov/data/china-sugar-annual-1

van der Ploeg, J. D., Franco, J. C., & Borras, S. M.Jr. (2015). Land concentration and land grabbing in Europe: A preliminary analysis. *Canadian Journal of Development Studies, 36*(2), 147–162.

van der Ploeg, J. D., Ye, J. Z., & Pan, L. (2014). Peasants, time and the land: The social organization of farming in China. *Journal of Rural Studies, 36*, 172–181.

Wei, F. A. (2014). *Yuenan Kanzhe Nvgong de Kuaguo Liudong Yanjiu* [On the transnational migration of the Vietnamese women labourers of sugarcane cutting in the Southwest Guangxi]. *Journal of Guangxi University for Nationalities (Philosophy and Social Science Edition), 36*(2), 72–77.

Wolford, W., Borras, S. M.Jr., Hall, R., Scoones, I., & White, B. (2013). Governing global land deals: The role of the state in the rush for land. *Development and Change, 44*(2), 189–210.

World Rainforest Movement. (2009). *Laos: Chinese company Sun Paper plans eucalyptus monocultures.* Retrieved from http://wrm.org.uy/articles-from-the-wrm-bulletin/section3/laos-chinese-company-sun-paper-plans-eucalyptus-monocultures/

Xu, Y. (in press-a). Politics of inclusion and exclusion in the Chinese industrial tree plantation sector: The global resource rush seen from inside China. *Journal of Peasant Studies.*

Xu, Y. (in press-b). Land grabbing involving foreign investors in China: Boom crops and rural conflict. *Third World Quarterly.*

Xu, Y. (in press-c). Loss or gain? Insights from Chinese villagers' livelihood change in the eucalyptus boom. *Canadian Journal of Development Studies.*

Yan, H. R., & Chen, Y. Y. (2015). Agrarian capitalization without capitalism? Capitalist dynamics from above and below in China. *Journal of Agrarian Change, 15*(3), 366–391.

Yan, H. R., Chen, Y. Y., & Ku, H. B. (2016). China's soybean crisis: The logic of modernization and its discontents. *The Journal of Peasant Studies, 43*(2), 373–395.

Ye, J. Z. (2015). Land transfer and the pursuit of agricultural modernization in China. *Journal of Agrarian Change, 15*(3), 314–337.

Ye, J. Z., Wu, H. F., Rao, J., Ding, B. Y., & Zhang, K. Y. (2016). Left-behind women: Gender exclusion and inequality in rural-urban migration in China. *The Journal of Peasant Studies, 43*(4), 910–941.

Yin, R., Xu, J., & Li, Z. (2003). Building institutions for markets: Experiences and lessons from China's rural forest sector. *Environment, Development and Sustainability, 5*(3), 333–351.

Zhang, H. Y., Liu, M., & Wang, H. (2002a). *Nongcun Tudi Shiyong Zhidu Bianqian: Jieduanxing, Duoyangxing yu Zhengcetiaozheng* [Land use institutional changes in rural China: Periodization, diversification and policy adjustments]. *Issues in Agricultural Economy, 2*, 12–20.

Zhang, H. Y., Liu, M., & Wang, H. (2002b). *Nongcun Tudi Shiyong Zhidu Bianqian: Jieduanxing, Duoyangxing yu Zhengcetiaozheng* [Land use institutional changes in rural China: Periodization, diversification and policy adjustments (second)]. *Issues in Agricultural Economy, 2*, 17–23.

Zhang, Q. F., Oya, C., & Ye, J. Z. (2015). Bringing agriculture back in: The central place of agrarian change in rural China studies. *Journal of Agrarian Change, 15*(3), 299–313.

Holding corporations from middle countries accountable for human rights violations: a case study of the Vietnamese company investment in Cambodia

Ratha Thuon

ABSTRACT

Land grabbing in poor countries by transnational corporations has been increasing, causing great concern over human rights violations in countries where states often lack the ability or will to regulate the conduct of foreign-owned companies. Civil society organizations have played a significant role in attempts to hold companies from Organisation for Economic Co-operation and Development (OECD) countries accountable for human rights violations by their subsidiaries in poor countries. However, civil society pressure for accountability from companies whose home base is in non-OECD, middle-income, countries is rare. This paper explores the human rights impacts of the Cambodian operations of Vietnam's Hoang Anh Gia Lai (HAGL) company, and how affected communities and NGOs in Cambodia have tried to hold HAGL accountable for its wrongdoing through approaching the Office of the Compliance Advisor Ombudsman of the International Finance Corporation. This is an initial attempt to examine how civil society and affected communities have challenged a Vietnamese company with no prior record of engaging with players from outside its home territory about the human rights impacts of its investments in Cambodia.

Introduction

Both investors from the Global North and investors from the South, including domestic elites, have become competitors with local farmers for land and natural resources across the world (Margulis & Porter, 2013). As the size and number of transnational land-based deals grow, so do attendant human rights abuses (Madeley, 2012) including forced displacement, child labour (Bendell, 2004), local economic impacts (Davis, D'Odorico, & Rulli, 2014), including acceleration of poverty and food insecurity (Demissie, 2015; Grant & Das, 2015), and destruction of natural resources and other environmental damage (Madeley, 2012). Some soft law and voluntary initiatives, for instance, the UN Global Compact and Organisation for Economic Co-operation and Development (OECD) Guidelines have been established and promoted by international organizations in order to place human rights obligations on transnational corporations (TNCs), as well as obligations on host and home states to protect people in host countries against human rights violations by corporations (Bunn, 2004). Advocates have argued that pursuing corporate responsibility through this kind of guideline initiative does not go far enough, and is ineffective mainly because of the voluntary nature and the lack of binding norms or mechanisms to ensure compliance (Bunn, 2004).

The conduct of TNCs and the limitations of the voluntary corporate social responsibility agenda has driven demands for corporate accountability (Utting, 2008). Giving someone responsibility to do something means 'it is up to them to take care of it' (Wenar, 2006, p. 5). In contrast, accountability 'implies some actors have the right to hold other actors to a set of standards, to judge whether they have fulfilled their responsibilities in light of these standards, and to impose sanction if they determine that these responsibilities have not been met' (Grant & Keohane, 2005, p. 29). For this reason, it is important that enforcement mechanisms have been put in place (Mulgan, 2000). The traditional notion of human rights accountability that focuses on state actors has recently been expanded to cover other actors such as TNCs, civil society organizations (CSOs), and international organizations (Putten, 2008). This paper focuses on corporate accountability.

How is it possible to hold corporations accountable? Private companies are chiefly accountable to shareholders or investors, but not necessarily to people affected by their operations (Koenig-Archibugi, 2004). Traditional international human rights law places primary obligations on host governments to prevent human rights abuses by TNCs. There is increasing recognition that home governments have an obligation to regulate the conduct of their TNCs in other countries. The United Nations Guiding Principles on Business and Human Rights has affirmed that, while states have the primary obligations under international human rights law, this does not absolve other parties, including business enterprises, of responsibility.[1] However, host states in the South often lack the ability or the political will to regulate TNCs and address the impacts of their business activities (Bunn, 2004). In addition, it is also extremely challenging for people negatively affected by the activities of TNCs to seek remedy for their grievances for reasons that include their lack of access to courts in the countries that have principal authority over the corporations in question (Hale, 2008).

Company shareholders and investors who profit from their operations are increasingly required to ensure that their money is not being invested in companies that cause harm to local communities and the environment (Newell & Bellour, 2002). CSOs have been pushing multilateral development banks (MDBs) to be more accountable since the late 1980s. They have demanded that such banks develop and implement environmental and social safeguard policies and accountability mechanisms (Ebrahim & Herz, 2007). The World Bank was the first MDB to respond to CSOs' calls for accountability as well as similar demands from the United States Congress by initiating an independent accountability mechanism (IAM) in 1993 called the Inspection Panel (Clark, 2003). The Inspection Panel provides an avenue for people affected by the World Bank-financed projects to file a complaint, and have their concerns investigated (Gualtieri, 2001). Following the example of the World Bank, other MDBs also established their own accountability mechanisms. IAMs contribute to accountability at MDBs, and they also promote accountability among their borrowers or clients (Gualtieri, 2001). For example, the Office of the Compliance Advisor Ombudsman (CAO) of the International Finance Corporation (IFC) has a problem-solving function, which mediates conflicts between affected communities and clients of the IFC, and a compliance review function (Compliance Advisor Ombudsman, 2015).

While approaching an IAM is a useful strategy, it is often not sufficient on its own to secure redress. Monitoring (Buntaine, 2015; Fukuda, 2003) and complementary CSO advocacy (Blackmore, Bugalski, & Pred, 2015) are important supplementary strategies. Examples of advocacy strategies used by CSOs include market pressure (Hale, 2008), naming and shaming (Utting, 2008) and lobbying (Marinettoa, 1998). Through market mechanisms, corporations can be pressured by those of their consumers who are concerned about environmental and social issues related to company activities, and by their investors who may refuse to invest in, or may divest from, companies whose activities cause negative environmental and social impacts (Hale, 2008). CSOs also play a crucial role by appealing to governments and corporations to address wrongdoing and exposing wrongdoing where necessary (Newell & Bellour, 2002). It should be noted that naming and shaming is an especially valuable advocacy strategy in respect of companies with well-known brand names (Klein, 1999) because they are more vulnerable to market and reputational damage (Grant & Keohane, 2005). Because corporations are chiefly accountable to their shareholders or investors (Koenig-Archibugi, 2004), CSOs may leverage the power of such actors for holding companies to account through soft advocacy strategies like lobbying (Marinettoa, 1998).

CSO trans-border advocacy campaigns that target corporations that hail from OECD countries are common and well known. However, there have been fewer CSO transnational campaigns that target investors from middle-income countries (Borras, Franco, & Wang, 2013). There are only a few cases where CSOs have targeted investors from middle-income countries for land-grabbing, particularly due to the lack of 'existing channels and rules of institutional interactions between them' (Borras et al., 2013, p. 65). Generally speaking, middle-income countries have weaker governance than OECD countries including regulatory quality, voice and accountability, rule of law, and control of corruption (Kaufmann, Kraay, & Mastruzzi, 2010). Rich countries also generally have more respect for human rights than low- and middle-income countries (Mitchell & McCormick, 1988). This makes it more difficult for CSOs to organize and mobilize in countries with weaker governance (Buntaine, 2015). Attempts to use judicial mechanisms in the South to hold corporations accountable have often been unsuccessful (Newell, 2001). In addition, while the effectiveness of CSOs' advocacy strategies depends very much on the reliability of information about 'where precisely to target advocacy campaigns' (Fox, 2007, p. 51) corporations operating in developing countries often lack commitment to transparency and disclosure (Graham & Woods, 2006).

This paper examines the human rights impacts of certain Cambodian operations of Hoang Anh Gia Lai (HAGL), a Vietnamese company, and how affected communities and supporting NGOs have been seeking redress from HAGL through the CAO for its wrongdoing in Cambodia. Information about HAGL's human rights violations and affected communities' access to local grievance mechanisms is based on the unpublished human rights impact assessment (HRIA) report, entitled 'Human rights impact assessment: Hoang Anh Gia Lai economic land concessions in Ratanakiri, Cambodia' written by Bugalski and Thuon in 2014. Information about CSOs' access to remedy through the CAO is based on the author's direct participant observation and personal working experience with affected communities and supporting NGOs in the HAGL case.

Methodology

Human rights that are acknowledged by legally binding and quasi-legal instruments provide a strong foundation for assessing a transnational investment's impacts on rights (Golay & Biglino, 2013). A number of legally binding treaties were used as the normative framework for analysing HAGL's human rights in Cambodia, including the International Covenant on Civil and Political Rights (ICCPR), the International Covenant on Economic, Social and Cultural Rights (ICESCR), the UN Convention on the Rights of the Child (CRC), and the United Nations Declaration on the Rights of Indigenous Peoples (UNDRIP). This paper focuses on the five most badly affected rights, namely the right to self-determination, the right to an adequate standard of living, the right to health, the right to enjoy culture and practice traditions, and the right to effective remedy. These five rights incorporate the main elements of the 11 'Minimum Principles' identified by UN Special Rapporteur on the right to food Olivier De Schutter to address the human rights issues of large-scale investment in farmland, namely the right to self-determination, the right to development, and the right to food

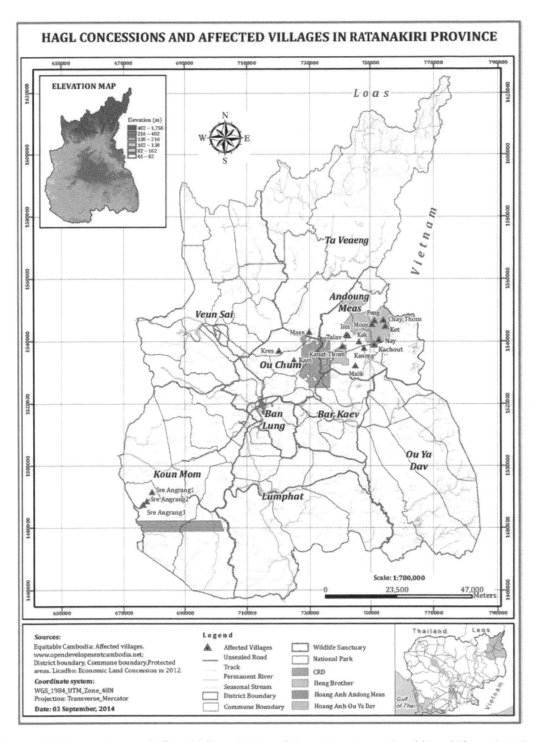

Figure 1. HAGL concessions and affected villages in Ratanakiri province. Source: Bugalski and Thuon (2014).

Table 1. HAGL's subsidiaries and affected villages.

HAGL subsidiaries	Original size of concession (ha)	Affected villages	Villages expected to be affected	District
Heng Brother Co. Ltd	2361	Kanat Thom Malik		Andong Meas
CRD Co. Ltd	7591	Kresh Kam Mass		O'Chum
Hoang Anh O'yadav Co. Ltd	9000	Inn Kak Mouy Peng Talao	Kachout Leur Tanong Nay Ket Chay	Andong Meas
Hoang Anh Andong Meas (Lumphat)	9470	Srae Angkrong 1 Srae Angkrong 2 Srae Angkrong 3		Koun Mom

Source: Open Development Cambodia; Global Witness (2013); correspondence with Global Witness for updated information; and interviews with village representatives.

(Claeys & Vanloqueren, 2013). The HRIA covers 18 villages or communities in Ratanakiri province of Cambodia that appear to be inside or near the boundary of HAGL concessions (Figure 1). Of the 18 villages, three are Khmer and the rest are occupied by indigenous groups, including Kachok, Jarai, Kreung, and Tampoun.

Thirteen of the 18 villages in this study have been affected by land and forest clearance by four subsidiaries of HAGL (see Table 1). The other five villages were reported to be located in the concession areas, but the company had not yet commenced its activities at the time of interview (November 2013 to March 2014).

Desk research was applied to gather background information on the concessions and villages as well as information on applicable laws and policies. The primary data was collected through key informant interviews, participatory community mappings, and group discussions, including separate women's group discussions with villagers in affected villages. The research team also interviewed 69 affected households. These tools and techniques were applied to collect information on impacts and losses that local people experienced and their efforts to seek redress at local and/or national level. The information on how affected villages and NGOs sought redress through the CAO process was drawn from participant observation and the author's personal working experience with affected villages and NGOs involved in the case.

Overview of economic land concessions (ELCs) and land disputes in Cambodia

The Royal Government of Cambodia (RGC) opened up private investment in the agriculture sector for both local and foreign investors in 1992 through granting ELCs to investors. According to the web page of the Ministry of Agriculture, Forestry and Fisheries (accessed 25 December 2015), the government's priority for agricultural sector investment was in large-scale plantation development on ELCs.[2] According to LICADHO Cambodia (2015), as at 23 October 2015, the RGC had granted 270 ELCs, a total of 2.1 million hectares of Cambodian land, to local and international companies. This data is incomplete and may not be accurate due to the government's lack of transparency. Vietnam has been leading the pack of TNCs that have been granted ELCs in Cambodia.[3] Most of the concessions owned by Vietnamese companies are concentrated in Ratanakiri and Kratie provinces.[4]

Despite the prevailing assumption that investment in agriculture in the form of large-scale agribusiness reduces poverty among local populations, ELCs are a major cause of land disputes, especially in rural areas (NGO Forum on Cambodia, 2015). Forced eviction in the context of land disputes is one of the most widespread and systematic human rights violations in Cambodia. According to Amnesty International, 150,000 people nationwide were facing the threat of forced eviction in 2008.[5] Land disputes are often related to unclear property rights and tenure insecurity as a legacy of civil war (Sperfeldt, Tek, & Chia-lung Tai, 2012).

There are five institutional structures that can be used to resolve land disputes, depending on whether the land is registered or not: (i) commune/sangkat councils, (ii) administrative committees; (iii) the Cadastral Commission; (iv) the National Authority for Land Dispute Resolution; and (v) the court system (Cambodian Center for Human Rights, 2014). The ability of these structures to resolve land disputes is hampered by implementation and enforcement difficulties, unclear areas of jurisdiction, time-consuming administrative processes, high costs associated with the submission of complaints, the lack of legal support, inconsistent decision-making, vulnerability to political pressure, and a judiciary that lacks independence. For these reasons, local people have no trust in referring land conflicts to these mechanisms for adjudication (Cambodian Center for Human Rights, 2014).

HAGL and the CAO process

HAGL is one of the top two Vietnamese rubber companies with investments in Cambodia. It was granted tens of thousands of hectares of land in Ratanakiri through its subsidiaries CRD, Heng Brother, Hoang Anh Andong Meas (Lumphat) and Hoang Anh O'Yadav. This makes HAGL one of the largest holders of ELCs in this province.[6] According to the Cambodian Land Law, the size of an ELC may not exceed 10,000 hectares. As a single legal entity, HAGL should be subject to this limit. Global Witness (2013) has identified Credit Suisse, Deutsche Bank, Dragon Capital Group and the IFC as HAGL investors. The IFC is indirectly exposed to HAGL through financial intermediaries Dragon Capital Group and VEIL after investing US$12 million in Dragon Capital/ VEIL in 2002 and US$8 million in 2003, according to relevant summary project information.

In 1999, the World Bank established the CAO as its internal mechanism to ensure compliance with IFC policies. People negatively affected by IFC-assisted projects can file a complaint with the CAO.[7] While parties to a complaint in the CAO problem-solving process are affected people or communities and the IFC's client, parties to the compliance review function are affected people or communities and the IFC itself (Office of the Compliance Advisor Ombudsman of the International Finance Corporation [CAO], 2015). If parties decide to start the problem-solving process, the CAO fosters mediation or negotiation to seek 'joint resolution' (CAO, 2015). An independent mediator will be contracted to facilitate the redress process. It is important to note that the problem-solving process is voluntary and, when it performs this function, the CAO does not identify mistakes or provide solutions to a problem. Parties have the right to officially withdraw themselves from the redress process at any time (CAO, 2015).

HAGL's human rights violations

Group discussions with villagers in affected villages revealed that, in most cases, no notice or information was provided before the company commenced operations, and when meetings did occur, the villages were not accurately or fully informed about the project or its impacts. In the few cases where documents and maps were presented, these were not in a form accessible to affected people. In addition, no effort was made to consult affected people about the decision of the government to grant the concessions or about HAGL's business activities, and no attempt was made by any actor to seek the free, prior and informed consent of affected people for projects with serious and direct effects on their lands and natural resources. This does not comply with the affirmation by the United Nations Expert Mechanism on the Rights of Indigenous Peoples that the UNDRIP requires the free, prior and informed consent of indigenous peoples to be obtained in matters of fundamental importance for their rights, survival, dignity and well-being.[8]

> Before [the concession], villagers could collect vegetables and other resources from the forest. Presently, the area where we could collect forest products is very much reduced because of the company's forest clearance activity. (A villager in Talao, 22 November 2013)

The company's operations have caused both household and communal losses in the 13 affected villages. Communal losses include the loss of collectively held and used lands and resources usually governed under a customary tenure system. Under Article 23 of the Cambodian Land Law, communities are entitled to interim protection that allows them to continue to manage their lands according to custom even before they are granted collective title. These include grazing land; land reserved for future generations and shifting cultivation; spirit forest, burial grounds or other sacred places; resin trees; water sources and fish resources; community forest; and access to state forest and the resources previously obtained there. These acts and omissions contravene Article 1 of the ICCPR and the ICESCR, as well as several articles of UNDRIP, including 26(2) and 32(2) that relate to the right to self-determination.

Losses of household property as a result of company land clearance include the loss of residential plots, rice fields and chamka (orchard/farming land), and crops. In some cases, the company destroyed houses and other shelters. Animals have also reportedly been killed or stolen by company workers. In many cases, households sold their land to the company under duress.

> We work very hard now … we do not have enough food to eat since the concession … when there is insufficient food we keep food for our kids and husband … it does not mean we do not eat, but we eat less than them. (Women's focus group discussion in Kak village, 19 November 2013)

Communal and household losses, including the loss of access to productive resources, have meant a fundamental deprivation of communities' means of subsistence. There has been a regression in the enjoyment of the right to an adequate standard of living among affected people. In particular, the loss of farming and grazing land, animals, fruit and vegetables sourced from the forest, and fish from streams has reduced the amount and variety of food resources available for household consumption. The loss of lands reserved for rotational agriculture poses a risk to future food security. The confiscation and destruction of these productive resources for present and future use has also led to a loss of sovereignty over food systems in affected villages, which people felt had previously provided them with healthy and culturally appropriate food in a sustainable manner. The cumulative effect of the loss of access to wild/ natural food and drop in income has meant that some people are facing difficulties in feeding their families, and the range of coping mechanisms has included increased debt. Should there be further loss of land and resources, there is a risk of serious food insecurity and deprivations of the right to food. The acts and omissions that caused these impacts amount to a violation of Article 11(1) of the ICESCR.

Communities losing spirit forests, burial grounds and other sacred sites has affected the enjoyment and practice of custom and culture of communities which amount to violations of Article 27 of the ICCPR, Article 30 of the CRC and Articles 11 and 12 of UNDRIP. In some villages, people have experienced regression in the enjoyment of the right to health that amount to a violation of Article 12 of the ICESCR. Adverse health impacts are mainly perceived to be due to pollution and destruction of the local environment and a resulting deterioration in the quantity and/or quality of food.

HAGL's compensation for losses

> First, I was threatened by police. Then there were three soldiers threatening me … They threatened me that whether I agree or disagree to give the land, it will be cleared anyway … One of the soldiers told me I should accept the cash compensation. (A villager in Kanat Thom, 17 November 2013)

Group discussions reveal that of approximately 164 households in the 13 affected villages that lost residential plots and/or individually held farmland to a HAGL subsidiary, 101 received cash compensation for seized land. The vast majority of households interviewed received less than US$300 in total. Many respondents believed that the amount of compensation they received was insufficient to make up for their losses. For them, cash—a short-term asset—cannot compensate for the loss of land. A total of 27 households in four villages accepted replacement plots offered by the company. At least five of these households claimed that the replacement land plots they received were smaller than the land plots that were taken from them. Thirty-six households in affected villages did not receive compensation of any kind for seized land. Only in a few cases did households report receiving compensation for lost crops. The company reportedly did not provide any compensation for structures that were destroyed or animals that were killed.

> The benefits the community received from the company were even smaller than a nail compared to the loss of land. (A villager in Kanat Thom, 17 November 2013)

Neither the company nor the government provided compensation to any of the affected villages for communal losses. In all but two villages, HAGL provided villagers with 'gifts' such as rice, salt and sugar, and small amounts of cash. Other contributions from the company included roads, wells, a community centre and medical examinations and assistance by a Vietnamese doctor. HAGL's medical programme has provided much-needed services to affected villages, with notable positive impacts for those who have received treatment for visual impairment and eye disease. The provision of regular ongoing free medical services is contributing to the progressive realization of the right to health.

The company states that it helps improve local livelihoods by providing job opportunities to local people. However, most people in affected villages do not work on HAGL plantations. Some villagers from Inn, Talao and Srae Angkrong 1, 2 and 3 have decided to work for the company as an alternative or supplementary source of income. Most of these are part-time or seasonal labourers, generally paid about US$6.25 per day to plant seedlings, weed, spread fertilizer and water the trees, among other tasks. Working hours are between 7:00 am and 5:00 pm, with a two-hour break. Children work on the plantation when they have free time. A few are reportedly under the age of 12, which is in breach of Cambodia's Labor Code and the International Labour Organization Minimum Age Convention.[9] Almost all respondents, including those who are ethnic Khmer, said that they prefer working on their own farms than the company's plantation. Most said working on their own farms is more convenient and flexible. Many also said they could derive greater benefits from their own farms. Other reasons include perceived difficult job conditions on the plantation, especially strict working hours and other rules, which they are not used to; not being comfortable working for someone else; and the perception that working on the plantation is not a sustainable source of income. The employment on rubber plantations of some people from affected villages has only partly mitigated income losses for affected households and has in no way compensated for lost control over food and livelihood sources and systems.

Although HAGL has undertaken some activities to demonstrate its corporate social responsibility as explained above, this does not mean it is showing any accountability for human rights violations caused by its subsidiaries. Eighty percent of households interviewed said they received benefits, but most of them said that these did not make up for the losses of land and natural resources that they had experienced. Some villagers described these benefits as gifts to persuade them not to lodge complaints against the company.

Access to remedy through local grievance mechanisms

The HRIA (Bugalski & Thuon, 2014) found obvious violations of human rights and of Cambodian law by HAGL operations. ELCs granted over indigenous communities' land or state public property are unlawful.[10] In Article 59, the Land Law also places limitations on the size of ELCs to mitigate the risk of the concentration of landholdings and the monopolization of arable lands. The company's failure to consult or negotiate with local residents also amounts to non-compliance with the requirements of Cambodian Sub-decree No. 146 and the concession agreements. Its forest clearance actions also violate Cambodian land and forestry laws. Local authorities have shown no willingness to hold HAGL to account.

One quarter of households interviewed said that they were threatened when they tried to get their land back. In Kanat Thom, after villagers confronted a bulldozer operator who was destroying their spirit forest, a policeman fired warning shots in their direction as they were returning to their village. Some said local authorities warned them not to make any complaints against the company. Villagers from Srae Angkrong 1, 2, and 3 were threatened with imprisonment by local authorities if they complained.

> To get our land back, we made countless complaints with thumbprints to local authorities. The complaint to commune office was rejected. Then, we submitted complaint to district office, but the district authority said they did not have any ability to resolve the problem. When our complaint reached provincial level, we were told that land was granted to the company and shown some legal document. (A villager in Srae Angkrong 3, 31 March 2014)

Despite the fear of retribution for expressing opposition to the HAGL project, most of the affected villages have submitted at least one petition and/or made at least one verbal complaint to local authorities, usually at the commune level. In some cases, villagers complained verbally to company workers. None of the key informants thought that their villages had received an adequate response to their complaints. Most of the complaints, both verbal and written, have been ignored. In other cases, complaints resulted in a 'take it or leave it' offer of compensation from the company, and the amount offered was perceived by affected people as inadequate.

Half of the respondents in household interviews had been involved in filing a complaint against the company. Of these respondents, about 40% said that after submitting the complaint they received cash compensation of an amount set by the company for individual household losses. In Srae Angkrong 1, 2, and 3, villagers verbally complained to the commune office, and Srae Angkrong 3 villagers also submitted several written complaints to commune and district offices. When they did not receive a response, they submitted a complaint through a local human rights NGO. As a result, the company provided cash compensation to households that had lost chamka and rice fields in these villages. Households who dared to negotiate with the company were offered US$200 per hectare. However, those who were less assertive were offered only US$100 per hectare. Affected households believed this compensation to be inadequate because it was less than the market price of the land taken by the company. Those who had not complained gave various reasons for this. One obstacle was limited education and knowledge about how to lodge a complaint, and with whom. Another obstacle is fear of the company, which people perceived as too rich and powerful to challenge. Having been informed by the government and the company that the concession is legal had also dissuaded some from protesting.

At the time of interview, affected people had not yet received effective remedies for the human rights violations they had experienced. While many villages and households had submitted complaints to local authorities and the company, these had either been ignored, met with threats, or addressed through inadequate offers of compensation without negotiation. No affected villages or households have attempted to bring a lawsuit through Cambodia's court system, despite having strong grounds under Cambodian law.

Access to remedy through the CAO

The RGCs' lack of ability or political will to address the negative impacts of HAGL activities has prompted affected people to seek other avenues of remedy. The lack of transparency about the land concessions, the beneficial owners of the plantations and other investors behind the project was one of the biggest challenges people from affected villages faced when seeking justice. In many cases, households in affected villages could not even properly name the HAGL subsidiaries that had been granted concessions.

Global Witness shared information about HAGL and its subsidiaries as well as its links to the IFC with a human rights organization, Inclusive Development International (IDI) to seek IDI's support for affected villagers to gain access to the IFC's CAO. In 2013, the Global Witness 'Rubber barons' report revealed that the IFC had indirectly invested in HAGL through a financial intermediary, Dragon Capital. The IFC was the first target for the advocacy campaign because it is the only one of HAGL's investors that has both safeguard policies and an accountability mechanism, the CAO, to which affected people can file a complaint. As IDI is an international NGO, it is difficult for its staff to reach and provide support to affected villages without assistance, so IDI sought to establish a partnership with a national human rights NGO, Equitable Cambodia (EC).

EC also found it difficult to work with affected villages because its office is located in Phnom Penh, the capital, about eight hours' drive from Ratanakiri. Most of the affected villagers are members of indigenous groups whose communication in Khmer is limited. Limited resources and skills meant that EC was not able to effectively support so many affected villages. For these reasons, EC and IDI built a network with local organizations, especially those working with indigenous people. In 2013 and 2014, a few consultation meetings were conducted with these local organizations to determine whether or not they wanted to get involved in the case. Three local organizations agreed to support the case, namely Highlander Association, Cambodia Indigenous Youth Association, and Indigenous Rights Active Members.

Meanwhile, a human rights impact assessment was conducted by EC and IDI to collect information and evidence of human rights violations. The information obtained through data collection was used for writing the complaint lodged with the CAO. The final report (Bugalski & Thuon, 2014) has been used as a basis for engaging with HAGL and its investors and other relevant actors in the conflict resolution process. The report has so far not been publicly released while the mediation process continues.

Accessing the CAO is a better option than doing nothing and waiting for the death. (A representative of Malik village, 6 February 2014)

To empower affected people in the decision-making process, NGOs informed them about HAGL, its investors and the CAO through workshops, meetings, pamphlets and a video animation in Khmer, which was also dubbed into four indigenous languages. In addition, EC organized at least two consultations to determine whether affected people wanted to file a complaint to the CAO for a problem-solving process. Fourteen of the 18 communities or villages included in the impact assessment decided to seek redress through the CAO. These 14 villages are affected by HAGL subsidiaries CRD, Heng Brother and Hoang Anh O'Yadav. Srae Angkrong 1, 2, and 3 (Khmer villages) and Chay were not parties to the complaint filed with the CAO.[11]

Affected villages submitted a complaint to the CAO with the support of the five NGOs in February 2014. The complaint highlighted IFC's financing of HAGL through a financial intermediary, VEIL/Dragon Capital. The communities' main reason for filing this complaint is to have their land returned to them. They are also seeking cash compensation for losses of crops, animals, structures and their investment in clearing and preparing land that was grabbed by the company. NGOs also sent the complaint to Dragon Capital, as an initial effort to lobby HAGL's investors to put pressure on HAGL through their business relationships. With the support of NGOs, affected people held a press conference to announce the submission of their complaint to the CAO with the aim of putting public pressure on HAGL and its investors to engage in the CAO's mediation process.

After the complaint was found eligible by the CAO, a mediator was appointed. Because the CAO problem-solving process is conducted on a voluntary basis, the mediator tried to bring the company to the negotiation table using soft techniques. Meanwhile, the HRIA was sent to HAGL's investors to inform them about how their money had been used to cause adverse impacts on local people and the environment and to seek their support for the conflict to be resolved. As a result, HAGL agreed to engage in the CAO process. However, the CAO process was blocked by the provincial governor of Ratanakiri in March 2014. IDI lobbied the World Bank and IFC in Washington DC to put pressure on the national government to support the CAO problem-solving process. Meanwhile, the CAO worked to get government approval for the process. The Ministry of Interior issued an official letter in October 2014 enabling the process to proceed after it had been blocked for about six months.

Because of the CAO's mediation process, HAGL issued two six-month moratoriums on clearance activities at three project areas in Cambodia: (i) Heng Brother; (ii) CRD; and (iii) Hoang Ahn O'Yadav. While this was significant for some villages, representatives of the most badly affected villages were not satisfied because they said it made no difference because the company had already cleared significant parts of their villages and forests. After an intensive series of negotiation meetings between the parties in September 2015, a major agreement was reached. HAGL undertook to stop clearing land in affected villages and to pay cash compensation or return the land to affected people if fact-finding trips found that the company had taken their land.[12] Returning the land to affected villages may not be as easy as what is written in the joint statement by parties. The state claims the concession area is state land that is being leased to the company, so only the government has authority to decide whether or not to give land to affected villagers. In addition, the company made a significant investment in developing its rubber plantations, and its commitment to returning land that has been developed is questionable. However, this agreement was the best affected villages could get from the negotiation meeting. This formal written agreement can be useful for villagers' advocacy campaign.

Fact-finding trips were conducted from January to July 2016. In April 2016, a joint meeting was conducted to discuss the results of previous fact-finding trips. The company informed CAO, NGOs and representatives of affected villages that the company had requested the government to reduce the size of its three concessions. A comparison of old and new concession maps shows that the total size of the three HAGL subsidiaries' concessions has been reduced from 18,952 hectares to 8371 hectares. Two villages, Kachout and Ket, are officially no longer inside the boundary of the Hoang Anh O'Yadav concession. However, the company requested the government to only exclude land from the concession that had not yet been converted into rubber plantations. NGOs sent letters to HAGL's investors to inform them about the situation and requesting them to put pressure on the company to provide redress to affected villages. So far, the NGOs have received no responses from the investors.

This may be the right time for NGOs and affected people to apply alternative advocacy strategies to hold HAGL to account. There are some cases of land disputes in Cambodia where affected communities have confronted abuses of human rights by engaging in resistance. For example, communities affected by the sugar concession in Kampong Speu province (Borras & Franco, 2013). The upcoming 2018 Cambodian national election can be a good opportunity for affected communities to seek political support from the government. The government is generally more responsive to community advocacy actions before a national election than afterwards. This does not necessary mean that affected communities have to walk away from the CAO problem-solving process. They could undertake advocacy activities in parallel with the CAO process, but should make sure that the two approaches support one another. Communities affected by the Phnom Penh Airport Expansion Project, for example, are known for undertaking protests in parallel with the CAO process, with a positive result.

Another option is filing a complaint with the court in Vietnam, but this will not be easy. In accordance with the United Nations Guiding Principles on Business and Human Rights, the Government of Vietnam has a duty to regulate HAGL's activities in Cambodia and elsewhere to improving its accountability and transparency, to prevent and redress human rights violations. However, the weak rule of law and the repressive political environment in Vietnam is the main reason why NGOs have never tried to pursue a legal case against HAGL in Vietnam. In addition, 'human rights organizations' are banned in Vietnam (Human Rights Watch, 2015). This makes it difficult for affected communities and NGOs in Cambodia to seek and gain full support from civil society in HAGL's home country.

Because corporations are more responsible for their investments than other actors (Koenig-Archibugi, 2004), it is important for NGOs to keep lobbying current and potential investors to use their leverage and influence in positive ways. Other advocacy strategies such as media advocacy, consumer advocacy and naming and shaming can be prepared. NGOs should keep themselves well-informed about HAGL's investment chain. Knowing current and potential key actors along the investment chain, their responsiveness to advocacy, and their ability to influence the company can help affected communities and NGOs to determine future advocacy targets and strategies, and determine which mechanisms they could use for seeking redress (Blackmore et al., 2015).

The CAO problem-solving process has yielded some positive impacts at the project level. This outcome was achieved through a multi-pronged sustained advocacy and mediation strategy. The strategy of lobbying current and potential investors of HAGL to keep the company negotiating in good faith through the CAO also played a significant role.

Conclusion

The confiscation of lands and destruction of forest resources by HAGL amount to the violations of local people's rights. However, none of those actors who make this project possible on the ground (governments of host and home countries and HAGL and its investors) have taken sufficient responsibility for their contribution to wrongdoing.

There is a lack of 'existing channels and rules of institutional interactions' between CSOs and investors from the middle-income countries (Borras et al., 2013). HAGL's projects in Cambodia are set within a national context of weak rule of law, systematic corruption and other poor governance factors making it difficult for affected people to seek remedies for human rights violations caused by its operations. Local grievance mechanisms in Cambodia fail to provide an effective remedy and a just solution for affected people. In addition, Vietnam is a lower middle-income country in which CSOs can find no reliable and effective means for holding HAGL accountable. HAGL is a giant Vietnamese rubber company that lacks safeguard policies or grievance mechanisms to address affected villages' concerns. Approaching the courts in Vietnam for remedy may be risky because it is a country where there is weak rule of law. In addition, the lack of recognition of human rights in Vietnam limits the space for CSOs in the host and home countries to effectively mobilize and seek accountability for human rights violations caused by HAGL operations in Cambodia.

Where TNCs have business relationships with international finance institutions that have safeguard policies and grievance mechanisms to ensure compliance with existing policies, there is space for CSOs to demand corporate accountability. Where an MDB is identified as an investor or financier, IAMs such as the CAO provide an avenue for affected people to lodge complaints and have their claims adjudicated. However, the CAO's problem-solving process through mediation is conducted on a voluntary basis. Although not all investors are responsive to CSO campaigns, complementary advocacy strategies such as lobbying with a range of investors is important for trying to keep HAGL engaging in the process in good faith. Because the CAO mediation process has not yet been concluded, it remains to be seen whether this is an effective mechanism for gaining redress. Demands for corporate accountability from companies from non-OECD countries that draw upon capital from MDBs can be made, at least to a certain extent. TNCs that do not want to take responsibility for their wrongdoings can try to escape from CSO pressure by seeking new sources of funding from the South or BRICs that are less responsive to their CSO advocacy than multilateral institutions.

Notes

1. Report of the Special Representative of the Secretary-General on the issue of human rights and transnational corporations and other business enterprises, John Ruggie, A/HRC/8/5, endorsed by the United Nations Human Rights Council in Resolution 17/4 of 16 June 2011.
2. For further details, see http://www.elc.maff.gov.kh/.
3. For further details, see LICADHO (2015).
4. Open Development Cambodia (2015).
5. Cambodia: Rights razed: Forced evictions in Cambodia, retrieved from http://www.refworld.org/docid/47b96a0e1a.html.
6. Global Witness (2013) and HAGL Group (2013).
7. For further details, see CAO (2013).
8. Report to the Human Rights Council of the Expert Mechanism on the Rights of Indigenous Peoples. A/HRC/EMRIP/2011/2, Annex: Expert Mechanism advice No. 2 (2011): Indigenous peoples and the right to participate in decision-making, paragraph 26.

9. Labor Code (1997), Article 177(4); and ILO Minimum Age Convention, 1973 (No. 138), Article 7 (ratified by Cambodia in 1999 and Vietnam in 2003).
10. See Land Law (2001), Article 15 and 58; and Sub-decree No. 146 (2005) on Economic Land Concessions, Article 4(1).
11. There were 17 villages that filed the complaint to the CAO. Three of them were not included in the impact assessment. HAGL confirmed that these three villages located outside its concession boundary, so they withdrew from the CAO mediation process.
12. For further details of the joint statement between HAGL and affected villages, see http://www.cao-ombudsman.org/cases/document-links/links-212.aspx.

Acknowledgements

I wish to express my sincere gratitude to Jun Borras for bringing together and framing this special issue; to Natalie Bugalski at Inclusive Development International for her critical and useful comment on an earlier version of this paper; to two peer reviewers for their very helpful and encouraging comments and suggestions to improve the paper's quality; and to staff at Equitable Cambodia for their contribution to the data collection of the Human Rights Impact Assessment.

Disclosure statement

No potential conflict of interest was reported by the author.

References

Amnesty International. (2008, February 11). *Cambodia: Rights razed: Forced evictions in Cambodia*. Retrieved from http://www.refworld.org/docid/47b96a0e1a.html
Bendell, J. (2004). *Barricades and boardrooms: A contemporary history of the corporate accountability movement*. Geneva: United Nations Research Institute for Social Development (Technology, Business and Society Programme Paper, no. 13).
Blackmore, E., Bugalski, N., & Pred, D. (2015). *Following the money: An advocate's guide to securing accountability in agricultural investments*. London/Asheville, NC: International Institute for Environment and Development/ Inclusive Development International.
Borras, S. M., & Franco, J. C. (2013). Global land grabbing and political reactions 'from below'. *Third World Quarterly, 34*(9), 1723–1747.
Borras, S. M., Franco, J. C., & Wang, C. (2013). The challenge of global governance of land grabbing: Changing international agricultural context and competing political views and strategies. *Globalizations, 10*(1), 161–179.
Bugalski, N., & Thuon, R. (2014). *Human rights impact assessment: Hoang Anh Gia Lai economic land concessions in Ratanakiri, Cambodia*. Unpublished report.
Bunn, I. D. (2004). Global advocacy for corporate accountability: Transatlantic perspectives from the NGO community. *American University International Law Review, 19*(6), 1265–1306.
Buntaine, M. T. (2015). Accountability in global governance: Civil society claims for environmental performance at the World Bank. *International Studies Quarterly, 59*(1), 99–111.
Cambodian Center for Human Rights. (2014). *The failure of land dispute resolution mechanisms* (Briefing Note – July 2014).

Cambodian Leage for the Promotion and Defense of Human Rights (LICADHO). (2015). *Cambodia's concessions*. Retrieved from www.licadho-cambodia.org/land_concessions.

Claeys, P., & Vanloqueren, G. (2013). The minimum human rights principles applicable to large-scale land acquisitions or leases. *Globalizations, 10*(1), 193–198.

Clark, D. (2003). Understanding the World Bank Inspection Panel. In D. Clark, F. Jonathan, & K. Treakle (Eds.), *Demanding accountability: Civil society claims and the World Bank Inspection Panel* (pp. 1–25). Lanham, MD: Rowman & Littlefield.

Davis, K. F., D'Odorico, P., & Rulli, M. C. (2014). Land grabbing: A preliminary quantification of economic impacts on rural livelihoods. *Population and Environment, 36*(2), 180–192.

Demissie, F. (2015). *Land grabbing in Africa: The race for Africa's rich farmland*. London: Routledge, Taylor & Francis Group.

Ebrahim, A., & Herz, S. (2007). *Accountability in complex organizations: World Bank responses to civil society*. Boston, MA: Harvard University Kennedy School of Government (Faculty Research Working Paper Working Paper; no. RWP07-060).

Expert Mechanism on the Rights of Indigenous Peoples. (2011). *Report to the UN Human Rights Council, A/HRC/EMRIP/2011/2, Annex: Expert Mechanism advice no. 2: Indigenous peoples and the right to participate in decision-making*.

Fox, J. (2007). *Accountability politics: Power and voice in rural Mexico*. New York, NY: Oxford University Press Inc.

Fukuda, K. (2003). Critical analysis of the new accountability mechanism of the Asian Development Bank. In D. Guerrero (Ed.), *Asienhaus, focus Asien: A handbook on the Asian Development Bank* (pp. 31–38). Essen: Asienstiftung.

Global Witness. (2013). *Rubber barons*. London: Global Witness.

Golay, C., & Biglino, I. (2013). Human rights responses to land grabbing: A right to food perspective. *Third World Quarterly, 34*(9), 1630–1650.

Graham, D., & Woods, N. (2006). Making corporate self-regulation effective in developing countries. *World Development, 34*(5), 868–883.

Grant, E., & Das, O. (2015). Land grabbing, sustainable development and human rights. *Transnational Environmental Law, 4*(2), 289–317.

Grant, R. W., & Keohane, R. O. (2005). Accountability and abuses of power in world politics. *American Political Science Review, 99*(1), 29–43.

Gualtieri, A. G. (2001). The environmental accountability of the World Bank to non-state actors: Insights from the Inspection Panel. *British Yearbook of International Law, 72*(1), 213–253.

Hale, T. N. (2008). Transparency, accountability, and global governance. *Global Governance, 14*(1), 73–94.

Hoang Anh Gia Lai (HAGL) Group. (2013). *Report of environmental and community development program in Laos and Cambodia*. Unpublished document.

Human Rights Watch. (2015). *World report 2015: Events of 2014*. New York, NY: Human Rights Watch.

Kaufmann, D., Kraay, A., & Mastruzzi, M. (2010). *The worldwide governance indicators: Methodology and analytical issues*. Washington, DC: World Bank (Policy Research Working Paper; no. 5430).

Klein, N. (1999). *No logo*. London: Flamingo.

Koenig-Archibugi, M. (2004). Transnational corporations and public accountability. *Government and Opposition, 39*(2), 234–259.

Labor Code of 1997 (Cambodia). Retrieved from http://www.ilo.org/dyn/travail/docs/701/Labour%20Law.pdf

Land Law of 2001 (Cambodia). Retrieved from http://www.cambodiainvestment.gov.kh/land-law_010430.html

Madeley, J. (2012). Land grabbing. *Appropriate Technology, 39*(2), 50–51.

Margulis, M. E., & Porter, T. (2013). Governing the global land grab: Multipolarity, ideas, and complexity in transnational governance. *Globalizations, 10*(1), 65–86.

Marinettoa, M. (1998). The shareholders strike back: Issues in the research of shareholder activism. *Environmental Politics, 7*(3), 125–133.

Mitchell, N. J., & McCormick, J. M. (1988). Economic and political explanations of human rights violations. *World Politics, 40*(4), 476–498.

Mulgan, R. (2000). 'Accountability': An ever-expanding concept? *Public Administration, 78*(3), 555–573.

Newell, P. (2001). Access to environmental justice? Litigation against TNCs in the south. *IDS Bulletin, 32*(1), 83–93.

Newell, P., & Bellour, S. (2002). *Mapping accountability: Origins, contexts and implications for development.* Brighton: Institute for Development Studies. (IDS Working Paper; no. 168.).

NGO Forum on Cambodia. (2015). *Statistical analysis of land disputes in Cambodia, 2014.* Phnom Penh: NGO Forum on Cambodia.

Office of the Compliance Advisor Ombudsman of the International Finance Corporation (CAO). (2013). *Operational guidelines.* Washington, DC: Author. Retrieved from www.cao-ombudsman.org/documents/ CAOOperationalGuidelines_2013.pdf.

Office of the Compliance Advisor Ombudsman of the International Finance Corporation (CAO). (2015). *2015 annual report.* Washington, DC: Author.

Open Development Cambodia. (2015). *Economic land concessions.* Retrieved from www.opendevelopment cambodia.net/company-profiles/economic-land-concessions.

Putten, M. V. (2008). *Policing the banks: Accountability mechanisms for the financial sector.* Kingston, ON: McGill-Queen's University Press.

Sperfeldt, C., Tek, F., & Chia-lung Tai, B. (2012). *An examination of policies promoting large-scale investments in farmland in Cambodia.* Phnom Penh: Cambodian Human Rights Action Committee.

Utting, P. (2008). The struggle for corporate accountability. *Development and Change, 39*(6), 959–975.

Wenar, L. (2006). Accountability in international development aid. *Ethics & International Affairs, 20*(1), 1–23.

Framing China's role in global land deal trends: why Southeast Asia is key

Elyse N. Mills

ABSTRACT

As Chinese investment in foreign land and agriculture expands dramatically worldwide, a growing body of research has emerged on the prevalence of land deals in Latin America and Africa. Southeast Asia, however, has only recently begun to receive significant attention in these discussions. A deeper exploration of the Southeast Asian context offers crucial insights into understanding the puzzle of global land deals (why, where, how they occur) more broadly. This paper frames this exploration via an overview of regional land deal trends – focusing particularly on China's emergence as a prolific investor in the Global South, and why Chinese investment is increasingly targeting Southeast Asia, especially vis-à-vis expanding flex and boom crop production. This paper aims to provide a broader contextualization for recent Southeast Asian case studies, and to highlight why more research in the region is key in deepening our understanding of global land deal trends.

Introduction

There has been a dramatic rise in foreign investment in agriculture and land in the past decade – particularly in the Global South. While in previous decades, agricultural investment mainly occurred in the food-processing sector, the recent convergence of multiple crises (food/fuel/finance/climate) has led to a surge in large-scale land deals, intended to increase exports of food and biofuel crops to investing countries (Food and Agriculture Organization of the United Nations [FAO], 2013). In the media, countries that have experienced rapid economic growth in the past few decades (e.g. China and Saudi Arabia) are publicized for their prominent roles in land deals worldwide (GRAIN, 2008). Yet, these reports have typically focused on investments in Africa and Latin America, which typically involve enormous tracts of land, while those occurring in Southeast Asia, which often involve smaller pieces of land, have received significantly less scholarly and media attention. Recently, however, this trend has begun to shift, as an increasing amount of research on foreign-owned land spikes emerges from Asia (see Castellanet & Diepart, 2015; Friis & Nielsen, 2016; Goetz, 2015).

China's unique experience with rapid economic and population growth, coupled with its limited natural resources and arable farmland, has garnered particular interest. Despite its remarkable food self-sufficiency campaigns, the recent expansion of its middle-class and subsequent changes in dietary patterns have contributed to a dramatic increase in demand for consumer goods – especially food. However, with 40 per cent of the world's farmers and only 9 per cent of the arable land, food security has become a critical concern for the Chinese government (GRAIN, 2008, p. 3). To cope with these constraints, the Chinese government has been pursuing land deals, both by engaging directly with foreign governments, as well as indirectly, by encouraging domestic companies to establish foreign partnerships (Borras & Franco, 2010). China's substantial foreign exchange reserves – which peaked at USD 3.8 trillion in 2014 – means it has valuable economic relationships, making the government's focus on foreign land much more feasible (GRAIN, 2008; World Bank, 2014a, p. 13).

In the Southeast Asian context, the rapid rise in land deals involving Chinese investors has only recently become a more heated topic of debate. This may be partly because land deals in the region used to be much less visible to the public, due to the lack of research and media coverage. The growth of many middle-income countries (MICs) in Southeast Asia (e.g. Malaysia, Indonesia, and Thailand), which has simultaneously drawn more international attention to the region and increased speculation over the value of its agricultural land, has contributed to the increasing visibility of land deals there. Foreign investors' interest in Southeast Asian land grew substantially between 2000 and 2012, as they recognized the potential of the region's emerging markets. Annual foreign investment in Cambodia and Laos grew 5–10 times (respectively) during that period (International Institute for Environment and Development [IIED], 2012). This rapid investment influx has contributed to a significant growth in the number of land deals occurring in Southeast Asia, with recent estimates showing that foreign investors have acquired millions of hectares of land since the late 1990s (IIED, 2012, p. 1).

This paper offers an overview of emerging global trends in Chinese investment in foreign agriculture and land. It highlights the importance of exploring the Southeast Asian context more deeply, arguing that this is crucial for both broadening our understanding of land deals, and contextualizing recent case studies on specific deals or countries in the region (see Friis & Nielsen, 2016; Goetz, 2015; Lamb & Dao, 2015). Thus, this paper first highlights key moments in China's investment trajectory, including its domestic limitations and global expansion. Second, it analyses emerging global trends, offering broad comparisons between Chinese land investments in Southeast Asia, Latin America, and Africa. Third, it explores two key factors that offer compelling insights into the intensification of Chinese investment in Southeast Asia – including regional trade relations and the increasing global demand for flex and boom crops. Finally, it offers some concluding remarks and questions for further research.

'Going global': offsetting domestic limitations

In just nine years (1999–2008), China rose from the world's 11th largest economy to the 3rd (following the European Union and the United States), with its GDP leaping from 1.6 to 7.1 per cent of the world total. Between the mid-1990s and 2012, its GDP grew more than 10 per cent per year, while that of high-income countries (those with a GNI of USD 12,746 or more) was less than 3 per cent during the same period (Nolan, 2012, p. 23; World Bank, 2014b). This was due in large part to the Chinese government's focus on developing sectors that produce labour-intensive industrial goods, taking advantage of the fact that the working-age population includes more than 950 million people – compared to 720 million people in all of the high-income countries combined (Nolan, 2012, p. 35).

While having a huge national workforce has contributed positively to economic growth, it also means there is an immense overall population (1.3 billion people) to provide for. This has raised questions regarding how China will continue to sustain its own growth and support its citizens without greater access to the required natural resources (Tilt, 2013). Not only will industrial emissions and waste continue to cause significant damage to its environment, its existing resources cannot regenerate fast enough to keep up with the intense demand. Thus, it is not surprising that Chinese investors are now looking beyond its borders and acquiring productive land in regions where natural resources are more plentiful. Since 2000, China has spent an estimated USD 115 billion on acquiring foreign land, which doubled its overseas investments from 25 to 50 billion in 2008 alone (Nolan, 2012, p. 10).

Chinese investment has mainly targeted resource-rich countries in the Global South, and has quickly penetrated Africa, Latin America, and Southeast Asia in the last few decades. China's interest in land investments abroad was first triggered in the post-World War II era, when it was involved in several development aid projects in Africa, intending to establish new political allies and a sense of solidarity with other 'Third World' countries. Many of these projects took the form of small-scale crop research farms that remained under local ownership, and as a result, foreign agricultural investment was viewed positively and strongly encouraged by African governments (Bräutigam, 2009; Hofman & Ho, 2012). By the 1990s, the global reach of Chinese investment began expanding, with 10 land deals (of a combined total of 11,000 hectares) reported (in online databases) in Africa, and 2 reported in Latin America, totalling several thousand hectares (exact number is unconfirmed). During the same period in Southeast Asia, six land deals in Cambodia alone made up a combined total of 105,000 hectares (Hofman & Ho, 2012, p. 13).

China's 1999 'Going Global Strategy', which centred on aid being 'mutually beneficial' for all parties involved, led to further expansion of Chinese economic involvement in the Global South by the early 2000s. The strategy's aim was for Chinese investors, in cooperation with local governments, to open up new land for development, establishing agricultural plantations, cultivating breeding technologies, and promoting the benefits of agricultural machinery and food processing – allegedly for the improvement of food security in the host countries. However, in the late 1990s, many Chinese farmers, who had become landless partly due to trade liberalization and rapid urbanization, were relocated to now Chinese-owned foreign farms, displacing local farmers and undermining their ability to provide food for their communities (Bräutigam & Tang, 2009).

The 'Going Global Strategy' facilitated a massive surge of Chinese investment in foreign agriculture in the last two decades, especially in Southeast Asia (see Figure 1). China is now one of the top three investors in Laos and Cambodia, accounting for 50 per cent of foreign capital investment in Laos' agricultural sector, and owning 50 per cent of foreign-owned land concessions in Cambodia (IIED, 2012, p. 3). This arguably makes China one of the most prolific investors in the region, which is reflective of broader trends in agricultural investment worldwide, in which China has emerged as a powerful player (IIED, 2012).

Taking the world by storm: trends in Chinese land investments

While Chinese investment in African and Latin American land has increasingly drawn attention from both scholars and the media in recent years, land deals occurring in Southeast Asia have received comparatively less attention. Considering the vast amount of Southeast Asian land being sold and leased to Chinese investors, this oversight significantly limits our understanding of broader land deal trends. For perspective, in all of Africa (54 countries), Chinese companies have invested in a total of 3.2 million hectares of land. Meanwhile, in Southeast Asia, they have invested in a total of 3.04 million hectares of land in just six countries (Indonesia, Papua New Guinea, the Philippines, Laos, Myanmar, and Cambodia) (Hofman & Ho, 2012, p. 17). This means that overall, Chinese investors may control more Southeast Asian land than in other regions, but there is a lack of sufficient data on land deals to know exactly how much.

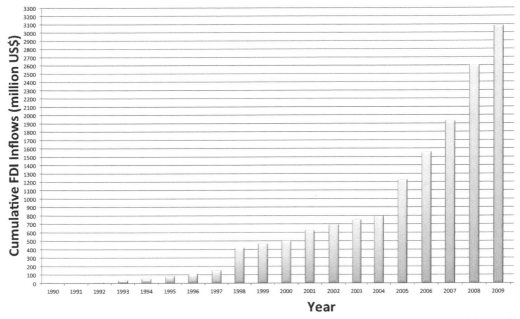

Figure 1. Chinese investment in Southeast Asia (1990–2009). Source: author's own construction (data collected from the Asian Development Bank database, 2014).

Certainly the limited nature of such statistics must not be overlooked. In fact, the incomplete picture we have of Southeast Asian land deals can be partly attributed to the limitations of databases, which often exclude data on 'smaller' land deals. The Land Matrix database, for instance, only reports land deals of 200 hectares or more (Land Matrix, 2016). This excludes many Southeast Asian deals, which *individually* tend to be smaller than those occurring in Africa or Latin America, but in *combination* constitute an enormous amount of land seized across the region.

The data that are available show that between 2000 and 2008, Chinese companies became particularly interested in Southeast Asia's Mekong region. In this area alone (namely Cambodia, Laos, and Myanmar), there were 25–30 investments reported (Bräutigam, 2009). In the post-2008 period, China began diversifying its geographical reach by pursuing land deals in regions that had not previously been targeted by other large-scale investors, such as Central Asia, Latin America, Eastern Europe, and the Pacific. Between 2009 and 2011, Chinese companies acquired around 35 tracts of land globally, with a combined total of approximately 2 million hectares. Interestingly, there were no Southeast Asian land deals recorded (in online databases), despite the 2010 establishment of the China-ASEAN Free Trade Area (Hofman & Ho, 2012, p. 18). Investment efforts also expanded into agriculture in highly industrialized countries, such as Australia, New Zealand, and parts of Europe. This reflected a renewed attempt on China's part to distance itself from the reputation it had garnered as a 'neo-colonialist' by establishing stronger trade relationships with 'stable' industrialized and emerging economies (Tan, 2014).

A 2013 FAO report shows that foreign investment in the Global South has been increasing at an average rate of 14.3 per cent in Latin America, 15.3 per cent in Africa, and 16.8 per cent in Asia since the 1990s. The 2007 investment spikes (illustrated in Figure 2) show that Africa received approximately USD 70 billion, Latin America peaked at 200 billion, while Asia reached a whopping 370 billion – five times the investment in Africa (FAO, 2013, pp. 16–18). Notably, these investment spikes coincide with the 2007–2008 world food price crisis, which significantly increased food insecurity in countries with large populations and little productive agricultural land – namely China. This intensified the urgency to secure resources beyond domestic borders.

A significant percentage of the foreign investment in Asia comes from public and private Chinese investors. Overall, China is the single largest investor in Asian agriculture – surpassed only by a *combined* total of investments from several smaller investing countries (see Figure 3). Chinese investments, both domestically and abroad, are characterized by careful state-led planning, intervention, and regulation. They also involve a complex web of public and private interests, making it difficult to determine exactly who is involved, and what factors are propelling a particular land deal (Hofman & Ho, 2012). These interests and factors vary depending on, for example, existing crop production systems in a region, or specific inter-country relations. In the Southeast Asian context, there are two key factors that stand out.

Flex and boom crops in your own backyard: targeting Southeast Asia

Two key factors offer compelling insights into the intensification of China's investment and involvement in Southeast Asia.

The first key factor is Southeast Asia's geographical proximity to China. Other regions that are experiencing an influx of land investment are usually conducting deals with distant countries (such as those between China and African countries) (see GRAIN, 2008, Annex). However, Chinese investors have also become increasingly interested in Southeast Asia, particularly as MICs, such as Malaysia, Indonesia, and Thailand, have seen significant economic growth in the past few years. This

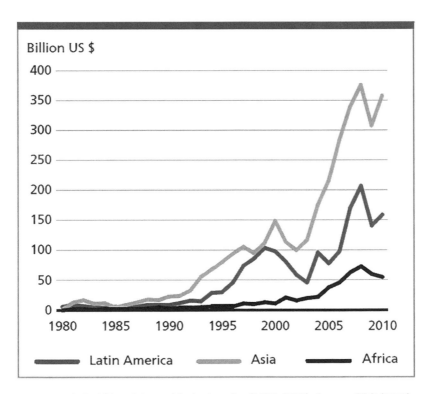

Figure 2. Investment trends in Africa, Asia, and Latin America (1980–2010). Source: FAO (2013).

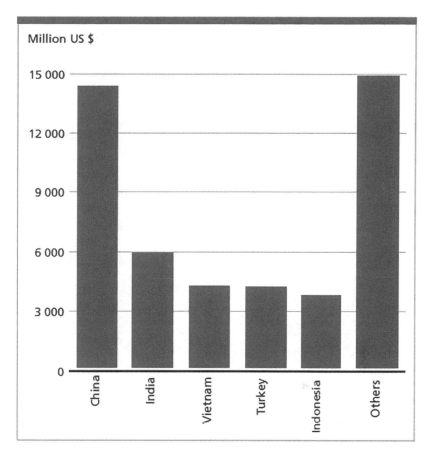

Figure 3. Agricultural investment into Asia (2003–2011). Source: FAO (2013).

has strengthened their position in the global economy, while also allowing for considerable expansion in their agricultural sectors, and thus offering a lucrative investment opportunity for foreign companies (Hall, 2009).

The Mekong region, where land is well irrigated, fertile, and contains vast areas of rice paddy (see Castellanet & Diepart, 2015), has drawn particular interest from Chinese investors. Between 2000 and 2008, China's presence in this area expanded significantly, with around 30 large-scale investments reported – mainly in cassava, palm oil, and sugar production (Hofman & Ho, 2012, pp. 15–16; IIED, 2012).

The Mekong region has also become an important area for Chinese investment because of existing historical trade routes, from southwestern China along the Mekong River, connecting to almost all of mainland Southeast Asia. These routes traditionally served as migration pathways out of China and into neighbouring countries (particularly Laos and Thailand), but the importance of this route has expanded considerably since the establishment of the Asian Development Bank's Greater Mekong Subregion (GMS) initiative in 1992. This programme intended to encourage economic cooperation via the 'Economic Quadrangle' between Thailand, Laos, Myanmar, and China's Yunnan Province. Part of what tightened this relationship was the construction of the North-South Economic Corridor (NSEC), a highway linking Kunming to Bangkok through the northern provinces of Laos (Tan, 2012; 2014).

The second key factor is the growing global demand for 'flex crops' and 'boom crops'. Flex crops are those that have multiple purposes (e.g. food, feed, fuel, and industrial products), such as corn, palm, soy, and sugarcane. Boom crops are high-value crops (e.g. flex crops), the production of which has suddenly skyrocketed due to intense demand (Borras, Franco, Isakson, Levidow, & Vervest, 2015). Since demand for flex and boom crops is increasingly the reason behind large-scale land deals, research on their production provides important insights into the dynamics and impacts of these deals. While Southeast Asian countries have experienced crop booms throughout history, those occurring in recent years differ because they almost exclusively involve export and high-value crop production (e.g. rubber, sugar, and lumber for biomass energy) (Hall, 2011; IIED, 2012). While these countries vary widely in terms of economic growth and development, they have all followed a common path toward predominantly export crop production (Hall, 2009). This path has also compelled governments in the region to further open their borders to foreign investors – particularly after the 2007–2008 world food price crisis.

The increasing demand for flex and boom crops also stems from the rapid rise in biofuel production. In Southeast Asia, Indonesian and Malaysian palm oil production dominates the biofuel industry, together accounting for 87 per cent of global production (IIED, 2012, p. 3). Tree plantations are also on the rise, with Cambodia and Indonesia devoting 60,000 and 200,000 hectares, respectively, to biomass production. Rubber production is even more significant, with Indonesia, Malaysia, and Thailand ranking as the top three producers in the world. Cambodia, Thailand, and Laos are also hotspots for producing sugarcane exports – with 80,000 hectares of Cambodian land being used for such projects (IIED, 2012, p. 3).

While policymakers justify this expansion for biofuel production using anti-fossil fuel and sustainable development narratives, analyses of the impact it has on environmental and labour standards are too often non-existent. Pro-biofuel arguments also fail to engage with discourse on changing power relations between transnational companies, states, and citizens. This is key to understanding the character of land deals, the role of the various actors involved, and the competition propelling agribusinesses to find the newest high-value crop or cheapest production methods (Borras et al., 2015). Some have argued for the need to more visibly embed discourses on land deals within analyses of contemporary capitalist development, highlighting that despite variations in the character of land deals (origin, destination, intention), they are all evidence of the crisis of the neoliberal globalization project (Hall, 2009). This crisis has emerged as a culmination of the food/fuel/finance/climate change crises occurring alongside the rise of the MICs, which require a continuous influx of resources in order to sustain their growth (2009).

Thus, these two key factors are useful framing points for deepening our understanding of the reasons behind the rapid increase in Chinese investment in Southeast Asia. They also provide useful contextualization for some of the emerging literature on crop production in the region, as researchers increasingly try to understand the dynamics surrounding *who is growing what and why* and how this is connected to changing regional power relations. This literature adds to debates on foreign land deals in three ways: it provides insights into *for whom* land becomes more valuable; it contributes detailed studies of *how* land control is actually implemented; and it strengthens policy debates around land deals (Hall, 2011).

Conclusions

This paper has provided an overview of emerging global trends in Chinese investment in foreign agriculture and land, highlighting the importance of exploring the Southeast Asian context more deeply. It has argued that this is crucial for both broadening our understanding of land deals and contextualizing recent cases studies in the region. It has also traced important historical moments in China's investment trajectory, offered broad comparisons between investments in Southeast Asia, Latin America, and Africa, and explored proximity and flex and boom crop production as key factors offering compelling insights into the intensification of Chinese investment in Southeast Asia.

Understanding the dynamics of such investments is crucial to understanding patterns of agrarian transformation in Southeast Asia, since these transactions contribute to transferring control of domestic markets into the hands of foreign companies (Liu, 2014). The Southeast Asian countryside is transforming, as smallholdings are converted into large-scale industrial plantations, transnational companies become agricultural producers, and farmers are dispossessed of their land and resources, excluded from markets, and forced to become sporadic labourers or factory workers (GRAIN, 2008). The combination of dispossession from land and resources, the presence of political actors (both in Southeast Asia and investing countries) that are allowing (or fuelling) this dispossession, and the increasing commodification of food and agriculture (Fairbairn et al., 2014) characterize such agrarian transformation. This significantly alters rural social and economic structures, as roles change within rural communities and power shifts out of the hands of people living and working on the land. As many emerging case studies on Southeast Asia reveal, such transformation has already had noticeable detrimental impacts (see Friis & Nielsen, 2016; Hall, 2011; Tan, 2014; among others).

This paper triggers several questions in relation to the land deal trends it has identified: (1) How are high-profile, large-scale land deals involving Chinese companies linked to everyday forms of land accumulation inside China and in Southeast Asia (as the latter are less visible and are thus less easily detected by formal regulatory institutions)? (2) How are Chinese investors competing or allying with investors from Southeast Asian MICs (such as Thailand and Vietnam), and with what implications? (3) How can international governance instruments be used to regulate land deals in the region, ensuring that no villagers are expelled from their lands, or are unjustly absorbed into agricultural companies? (4) In such complicated political terrain, how will civil society organizations (CSOs) carry out advocacy work when the target – China – is not accustomed to CSO engagement? These are some key questions that should be explored in future research, focusing on both the potential long-term implications of regional trends and their broader global significance.

Acknowledgements

I would like to thank the editors for all of their hard work in developing this exciting Special Issue and for giving me the opportunity to have my article included in the issue. And I would like to extend my sincere appreciation to anonymous reviewers for their thorough and thoughtful comments, which have helped significantly in sharpening and clarifying the arguments in this paper.

Disclosure statement

No potential conflict of interest was reported by the author.

Funding

I would like to thank the Initiatives in Critical Agrarian Studies (ICAS) for the funding support for this research.

References

Asian Development Bank: Asian Regional Integration Center. (2014, June). Retrieved from http://aric.adb.org/integrationindicators

Borras, S., & Franco, J. (2010). From threat to opportunity? Problems with the idea of a 'code of conduct' for land-grabbing. *Yale Human Rights & Development Law Journal, 13*(2), 507–523.

Borras, S., Franco, J., Isakson, R., Levidow, L., & Vervest, P. (2015). The rise of flex crops and commodities: Implications for research. *The Journal of Peasant Studies, 43*(1), 93–115.

Bräutigam, D. (2009). *The dragon's gift: The real story of China in Africa.* New York, NY: Oxford University Press.

Bräutigam, D., & Tang, X. (2009). China's engagement in African agriculture: 'Down to the countryside'. *The China Quarterly, 199,* 686–706.

Castellanet, C., & Diepart, J. C. (2015, June 5–6). *The neoliberal agricultural modernization model: A fundamental cause for large-scale land acquisition and counter land reform policies in the Mekong Region.* Conference paper No. 55, Land grabbing, conflict and agrarian-environmental transformations: Perspectives from East and Southeast Asia. Chiang Mai: Chiang Mai University.

Fairbairn, M., Fox, J., Ryan Isakson, S., Levien, M., Peluso, N., Razavi, S., … Sivaramakrishnan, K. (2014). Introduction: New directions in agrarian political economy. *The Journal of Peasant Studies, 41*(5), 653–666.

Food and Agriculture Organization of the United Nations. (2013). *Trends and impacts of foreign investment in developing country agriculture: Evidence from case studies.* Rome: Author.

Friis, C., & Nielsen, J. (2016). Small-scale land acquisitions, large-scale implications: Exploring the case of Chinese banana investments in Northern Laos. *Land Use Policy, 57,* 117–129.

Goetz, A. (2015, June 5–6). *Different regions, different reasons? Comparing Chinese land-consuming outward investments in Southeast Asia and Sub-Saharan Africa.* Conference paper No. 37, Land grabbing, conflict and agrarian-environmental transformations: Perspectives from East and Southeast Asia. Chiang Mai: Chiang Mai University.

GRAIN. (2008). *Seized: The 2008 land grab for food and financial security.* Barcelona: Author.

Hall, D. (2009). The 2008 world development report and the political economy of Southeast Asian agriculture. *The Journal of Peasant Studies, 36*(3), 603–609.

Hall, D. (2011). Land grabs, land control, and the Southeast Asian crop booms. *The Journal of Peasant Studies, 38*(4), 837–857.

Hofman, I., & Ho, P. (2012). China's 'developmental outsourcing': A critical examination of Chinese global 'land grabs' discourse. *The Journal of Peasant Studies, 39*(1), 1–48.

International Institute for Environment and Development. (2012). *Agricultural land acquisitions: A lens on Southeast Asia, April 2012. Briefing: The global land rush.* London: Author.

Lamb, V., & Dao, N. (2015, June 5–6). *Perceptions and practices of investment: China's hydropower investments in Mainland Southeast Asia.* Conference paper No. 21, Land grabbing, conflict and agrarian-environmental transformations: Perspectives from East and Southeast Asia. Chiang Mai: Chiang Mai University.

Land Matrix. (2016, March). Retrieved from http://www.landmatrix.org/en/

Liu, P. (2014). *Impacts of foreign agricultural investment on developing countries: Evidence from case studies* (FAO Commodity and Trade Policy Research Working Paper No. 47). Rome: FAO.

Nolan, P. (2012). *Is China buying the world?* Cambridge: Polity Press.

Tan, D. (2012). Small is beautiful: Lessons from Laos for the study of Chinese overseas. *Journal of Current Chinese Affairs, 41*(2), 61–94.

Tan, D. (2014). *China in Laos: Is there cause for worry?* (ISEAS Perspective #31). Singapore: Institute of Southeast Asian Studies (ISEAS).

Tilt, B. (2013). The politics of industrial pollution in rural China. *The Journal of Peasant Studies*, *40*(6), 1147–1164.

World Bank. (2014a). *China economic update June 2014. Special topic: Changing food patterns in China: Implications for domestic supply and international trade.* Washington, DC: World Bank Group.

World Bank. (2014b, September). Countries and Economies Online Data. Retrieved from http://data.worldbank.org/country

Index

Note: page numbers in italics and bold refers to figures and tables, page numbers followed by 'n' refers to endnotes.